Lecture Notes in Mathematics

Edited by A. Dold and B. Eckmann

1113

Piotr Antosik
Charles Swartz

Matrix Methods in Analysis

Springer-Verlag
Berlin Heidelberg New York Tokyo 1985

Authors

Piotr Antosik
Institute of Mathematics, Polish Academy of Sciences
ul. Wieczorka 8, 40-013 Katowice, Poland

Charles Swartz
Department of Mathematical Sciences, New Mexico State University
Las Cruces, N.M. 88003, USA

AMS Subject Classification (1980): Primary: 46-0, Secondary: 28 C 20,
40 A 05, 40 C 05, 46 A 15, 46 G 10

ISBN 3-540-15185-0 Springer-Verlag Berlin Heidelberg New York Tokyo
ISBN 0-387-15185-0 Springer-Verlag New York Heidelberg Berlin Tokyo

Printing and binding: Beltz Offsetdruck, Hemsbach/Bergstr.
2146/3140-543210

Contents

1. Introduction

In this set of lecture notes, we present a culmination of results on infinite matrices which were evolved by the members of the Katowice Branch of the Mathematics Institute of the Polish Academy of Sciences. In the early history of functional analysis "sliding hump" methods were used extensively to establish some of the early abstract results in functional analysis. For example, the first proofs of versions of the Uniform Boundedness Principle by Hahn and Banach and Hildebrand utilized sliding hump methods ([18], [39], [42], [35]). Since Banach and Steinhaus gave a proof of the Uniform Boundedness Principle based on the Baire Category Theorem, category methods have proven to be very popular in treating various topics in functional analysis [19]. In recent times, there has been a return to "sliding hump" methods in treating various topics in functional analysis and measure theory. For example, in [34] Diestel and Uhl use a lemma of Rosenthal ([64]) as an abstract sliding hump method to treat a variety of topics in vector measure theory.

In a somewhat similar fashion, the Antosik-Mikusinski Diagonal Theorem ([53], [2], [3], [9]) can be considered to be an abstract sliding hump method and has been employed to treat a wide variety of topics in functional analysis and measure theory ([4], [5], [6], [9], [12], [53], [54], [56], [57]). The Antosik-Mikusinski Diagonal Theorem is a result concerning infinite matrices and has proven to be quite effective in treating various topics that were previously treated by Baire category methods (see in particular the texts [12], [56]). These notes present a result concerning infinite matrices which is of an even simpler and more elementary character than the Diagonal Theorem, and which can still be used to treat a wide variety of topics in functional analysis and measure theory ([16]).

In section 2, we present the two basic matrix results evolved by P. Antosik in references [6] - [11], and then in subsequent sections we present various applications of the matrix results to topics in functional analysis and measure theory. After the basic material has been presented in sections 2 and 3, there has been an attempt to make the subsequent chapters on applications independent of one another. Thus, there is some repetition in some of the chapters; for example, summability is mentioned in both sections 5 and 8 and other topics are repeated.

In section 3, we introduce and study the notions of $\mathcal{K}$ convergence and $\mathcal{K}$ boundedness which were also discovered and studied by the Katowice mathematicians ([6] - [11]). An equivalent form of $\mathcal{K}$ convergence was introduced by S. Mazur and W. Orlicz in [52] and also studied by A. Alexiewicz in [1]. The idea was rediscovered in the seminar of P. Antosik and J. Mikusinski. In subsequent sections the notions of $\mathcal{K}$ convergence and $\mathcal{K}$ boundedness will be shown to be effective substitutes for completeness and barrelledness assumptions in many of the classical results of functional analysis. For example, in section 4, we treat the Uniform Boundedness Principle. The classical Uniform Boundedness Principle is well-known to be false in the absence of completeness or barrelledness assumptions, but we present a version of the Uniform Boundedness Principle in Theorem 4.2 which is valid in the absence of any completeness assumption and which contains the classical Uniform Boundedness Theorem for F-spaces as a special case. To illustrate the utility of our general Uniform Boundedness Principle in the absence of completeness, we give a derivation of the Nikodym Boundedness Theorem based on the general Uniform Boundedness Principle.

In section 5, we discuss a classical result on the convergence of operators which is sometimes attributed to Banach and Steinhaus. This result, like the Uniform Boundedness Principle, is known to be

false without completeness or barrelledness assumptions. Neverthe-
less, using the notion of $\mathcal{K}$ convergence, we present a version of
this theorem which is valid without any completeness assumptions. As
an application of the general result in the absence of completeness,
we use it to derive the Nikodym Convergence Theorem, the Brooks-
Jewett Theorem, and a result of Hahn, Schur and Toeplitz on
summability.

In section 6, we treat bilinear maps using our matrix methods.
We derive the classical result of Mazur and Orlicz on the joint
continuity of separately continuous bilinear maps and also, using the
notion of $\mathcal{K}$ convergence, present several hypocontinuity type of
results which are valid without completeness assumptions. Our
hypocontinuity results generalize results of Bourbaki.

In section 7, we treat various Orlicz-Pettis type results on
subseries convergent series by matrix methods. We derive the
classical Orlicz-Pettis Theorem as well as Orlicz-Pettis results for
compact operators and the topology of pointwise convergence on
certain well-known function spaces.

In section 8, we give generalizations of the classical lemmas of
Schur and Phillips to the group-valued case. We show that these general results contain the classical lemmas of Schur and Phillips as
special cases. A result of Hahn and Schur on summability is also
obtained from the general results.

In section 9, we present a version of the Schur lemma for
bounded multiplier convergent series in a metric linear space. This
version for bounded multiplier convergent series is motivated by a
sharper conclusion of the classical Schur lemma for B-spaces which
is obtained in Corollary 8.4. Some general remarks on the vector
versions of the summability results of Schur and Hahn are also
included.

In section 10, we consider the problem of imbedding c_o and ℓ^∞ into a B-space. Using the basic matrix lemma of section 2, we obtain the classical results of Bessaga-Pelczynski and Diestel-Faires on imbedding c_o and ℓ^∞ into B-spaces. We also indicate applications to a large number of well-known results in Banach space theory. The results and method of proof are very analogous to those of Diestel and Uhl ([34] I.4) except that the basic matrix lemma is employed instead of the Rosenthal lemma.

There are two themes which prevail throughout these notes. The first is that the matrix results presented here, although being very elementary in character, are extremely effective in treating various topics in measure theory and functional analysis which have been traditionally treated by Baire category methods. The other theme is that the idea of $\mathcal{K}$ convergence can be used as an effective substitute for completeness assumptions in many classical results in functional analysis. For example, we present versions of the Uniform Boundedness Principle, the Banach-Steinhaus Theorem and classical hypocontinuity results which are valid with no completeness assumptions whatever being present. Applications of these general results in the absence of completeness are indicated.

Many of the topics treated in these notes are standard topics in functional analysis which are treated in a great number of the functional analysis texts by various means including the popular Baire category methods. The matrix methods employed in these notes are of a very elementary character and can be presented without requiring a great deal of mathematical background on the part of the reader. For this reason these matrix methods would seem to be quite appropriate for presentation of some of the classical functional analysis topics to readers with modest mathematical backgrounds. It is the authors' hope that the matrix methods presented here will find their way into the future functional analysis texts.

We conclude this introduction by fixing the notation which will be used in the sequel.

Throughout the remainder of these notes, unless explicitly stated otherwise, E, F and G will denote normed groups. That is, E is assumed to be an Abelian topological group whose topology is generated by a quasi-norm $|\,| : E \to \mathbf{R}_+$. ($|\,|$ is a quasi-norm if $|0| = 0$, $|-x| = |x|$ and $|x+y| \leqslant |x| + |y|$; a quasi-norm generates a metric topology on E via the translation invariant metric $d(x,y) = |x-y|$.)

Recall that the topology of any topological group is always generated by a family of quasi-norms ([27]). Thus, many of the results are actually valid for arbitrary topological groups. We present the results for normed groups only for the sake of simplicity of exposition.

Similarly, X, Y and Z will denote metric linear spaces whose topologies are generated by a quasi-norm $|\,|$. (For convenience, all vector spaces will be assumed to be real; most of the results are valid for complex vector spaces with obvious modifications.) If it is further assumed that X is a normed space, we write $|\,|\,|\,|$ for the norm on X.

The space of all continuous linear operators from X into Y will be denoted by L(X,Y). If X and Y are normed spaces, the operator norm of an element $T \in L(X,Y)$ is defined by $||T|| = \sup\{||Tx|| : ||x|| \leqslant 1\}$.

If X and Y are two vector spaces in duality with one another by the bilinear pairing $<,>$, the weakest topology on X such that the linear maps $x \to <x, y>$ are continuous for all $y \in Y$ is denoted by $\sigma(X,Y)$. $\sigma(X,Y)$ is referred to as the weak topology on X induced by Y ([79] 8.2).

Other notations and terminology employed in the notes is standard. Specifically, we follow [38] for the most part.

Finally, for later use, we record a lemma of Drewnowski ([36]) which will be used at several junctures in the text.

Let Σ be an algebra of subsets of a set S. If $\mu : \Sigma \to G$ is a finitely additive set function, then μ is said to be <u>strongly additive</u> (<u>exhaustive</u> or <u>strongly bounded</u>) if $\lim \mu(E_i) = 0$ for each disjoint sequence $\{E_i\}$ from Σ. We have the following result due to Drewnowski.

<u>Lemma 1</u>. Let Σ be a σ-algebra. If $\mu_i : \Sigma \to G$ is a sequence of strongly additive set functions and $\{E_j\}$ is a disjoint sequence from Σ, then there is a subsequence $\{E_{k_j}\}$ of $\{E_j\}$ such that μ_i is countably additive on the σ-algebra generated by $\{E_{k_j}\}$.

Drewnowski states this result for a single strongly additive measure in [36] (see also Diestel and Uhl [34] I.6), but the lemma above can be derived from Drewnowski's result in the following way: let G^N be the space of all G-valued sequences. Equip G^N with the quasi-norm $| \ |$ defined by

$$|g| = \sum_{i=1}^{\infty} |g_i| / (1 + |g_i|) 2^i$$

where $g = (g_1, g_2, \ldots)$ and $|g_i|$ is the "norm" of g_i in G. Now define $\mu : \Sigma \to G^N$ by $\mu(E) = (\mu_1(E), \mu_2(E), \ldots)$. Then μ is strongly additive so by Drewnowski's lemma, there is a subsequence $\{E_{k_j}\}$ of $\{E_j\}$ such that μ is countably additive on the σ-algebra, Σ_0, generated by $\{E_{k_j}\}$. Then each μ_i is clearly countably additive on the σ-algebra Σ_0.

2. Basic Matrix Results

In this section we establish the two basic results on infinite matrices which will be used throughout the sequel. The first result is a very simple and elementary result on matrices of non-negative real numbers. This result is then used to establish a convergence type result for matrices with elements in a topological group. Both results are of an elementary character and require only elementary techniques in their proofs.

Lemma 1. Let $x_{ij} \geqslant 0$ and $\epsilon_{ij} > 0$ for $i, j \in \mathbf{N}$. If $\lim_i x_{ij} = 0$ for each j and $\lim_j x_{ij} = 0$ for each i, then there is a subsequence $\{m_i\}$ of positive integers such that $x_{m_i m_j} < \epsilon_{ij}$ for $i \neq j$.

Proof: Put $m_1 = 1$. There is an $m_2 > m_1$ such that $x_{m_1 m} < \epsilon_{12}$ and $x_{mm_1} < \epsilon_{21}$ for $m \geqslant m_2$. Then there is an $m_3 > m_2$ such that $x_{m_1 m} < \epsilon_{13}$, $x_{m_2 m} < \epsilon_{23}$, $x_{mm_1} < \epsilon_{31}$ and $x_{mm_2} < \epsilon_{32}$ for $m \geqslant m_3$. An easy induction completes the proof.

Lemma 1 will be used directly in several later results but the principle application of Lemma 1 will be to establish the basic matrix convergence result below.

Basic Matrix Theorem 2. Let E be a normed group and $x_{ij} \in E$ for $i, j \in \mathbf{N}$. Suppose

(I) $\lim_i x_{ij} = x_j$ exists for each j and

(II)   for each subsequence $\{m_j\}$ there is a subsequence $\{n_j\}$ of

$\{m_j\}$ such that $\{\sum_{j=1}^{\infty} x_{i n_j}\}$ is Cauchy.

Then $\lim_i x_{ij} = x_j$ uniformly with respect to j.

In particular, $\lim_i x_{ii} = 0$.

Proof: If the conclusion fails, there is a $\delta > 0$ and a subsequence $\{k_i\}$ such that $\sup_j |x_{k_i j} - x_j| > \delta$. For notational convenience assume $k_i = i$. Set $i_1 = 1$ and pick j_1 such that $|x_{i_1 j_1} - x_{j_1}| > \delta$.

By (I) there is $i_2 > i_1$ such that $|x_{i_1 j_1} - x_{i_2 j_1}| > \delta$ and

$|x_{ij} - x_j| < \delta$ for $i > i_2$ and $1 \leqslant j \leqslant j_1$. Now pick j_2 such that

$|x_{i_2 j_2} - x_{j_2}| > \delta$ and note that $j_2 > j_1$. Continuing by induction,

we obtain subsequences $\{i_k\}$ and $\{j_k\}$ such that

$|x_{i_k j_k} - x_{i_{k+1} j_k}| > \delta$. Set $Z_{k\ell} = x_{i_k j_\ell} - x_{i_{k+1} j_\ell}$ and note

(1) $$|Z_{kk}| > \delta.$$

Consider the matrix $[|Z_{k\ell}|] = Z$. By (I), the columns of this matrix converge to 0. By (II), the rows of the matrix $[x_{ij}]$ converge to 0 so the same holds for the matrix Z. Let $\epsilon_{ij} > 0$ be such that $\sum_{ij} \epsilon_{ij} < \infty$. By Lemma 1, there is a subsequence $\{m_k\}$ such that $|Z_{m_k m_\ell}| < \epsilon_{k\ell}$ for $k \neq \ell$.

By (II) there is a subsequence $\{n_k\}$ of $\{m_k\}$ such that

(2) $$\lim \sum_{\ell=1}^{\infty} Z_{n_k n_\ell} = 0.$$

Then

(3) $$|Z_{n_k n_k}| \leqslant |\sum_{\ell \neq k} Z_{n_k n_\ell}| + |\sum_{\ell=1}^{\infty} Z_{n_k n_\ell}| < \sum_{\ell \neq k} \epsilon_{n_k n_\ell} +$$

$$\left| \sum_{\ell=1}^{\infty} z_{n_k n_\ell} \right|.$$

Now the first term on the right hand side of (3) goes to 0 as $k \to \infty$ by the convergence of the series $\Sigma \, \epsilon_{k\ell}$ and the second term goes to 0 by (2). But this contradicts (1) and establishes the first part of the conclusion.

The uniform convergence of the limit, $\lim_i x_{ij} = x_j$ and the fact that $\lim_j x_{ij} = 0$ for each i implies that the double limit $\lim_{ij} x_{ij}$ exists and is equal to 0. In particular, this implies $\lim_i x_{ii} = 0$.

This matrix result will be the basic tool used throughout the sequel. A matrix $[x_{ij}]$ which satisfies conditions (I) and (II) of Theorem 2 will be called a $\mathfrak{K}$ matrix (the reason for the use of this terminology will be indicated in the next section).

The Basic Matrix Theorem 2 has a very different character than the Antosik-Mikusinski Diagonal Theorem in that the hypothesis and the conclusions have very different forms ([2], [53]). However, Theorem 2 can also be viewed as a diagonal theorem in the sense that the hypotheses of Theorem 2 imply that the diagonal sequence $\{x_{ii}\}$ converges to zero. In fact, if one first shows only that the diagonal sequence converges to zero, then it is not difficult to use this to show that in fact the columns of the matrix are uniformly convergent.

Matrix results of a very similar nature to Theorem 2 have been established in [6] - [11] and [73]. The matrix results of these papers have been used to treat a wide variety of topics in both measure theory and functional analysis. Much of the content of these papers will be treated in chapters 4, 5, 8 and 9.

It should be pointed out that the functional analysis text of

E. Pap ([56]) uses the Antosik-Mikusinski Diagonal Theorem in a systematic manner to treat a variety of classical topics in functional analysis and in this sense is very much in the spirit of these notes except that we systematically employ the Basic Matrix Lemma.

3. $\mathcal{K}$ Convergence

In this section we introduce the notion of a $\mathcal{K}$ convergent sequence. This notion was introduced by P. Antosik in [6] and was further explored in [7] - [11]; further applications to the Uniform Boundedness Principle and bilinear maps are given in [14] and [75]. The " $\mathcal{K}$ " in the description below is in honor of the members of the Katowice Branch of the Mathematics Institute of the Polish Academy of Sciences who have extensively studied and developed many of the results pertaining to $\mathcal{K}$ convergent sequences.

As a historical note, it should be pointed out that S. Mazur and W. Orlicz introduced a concept very closely related to that of a $\mathcal{K}$ convergent sequence in [52], Axiom II, p. 169. They essentially introduced the notion of a $\mathcal{K}$ (metric linear) space which is defined below and noted that the classical Uniform Boundedness Principle holds in such spaces. A. Alexiewicz also studied consequences of this notion in convergence spaces ([1] axiom A_2' , p. 203). It should also be noted however that the notion of a $\mathcal{K}$ convergent sequence and that of a $\mathcal{K}$ bounded set permits the formulation of versions of the Uniform Boundedness Principle in arbitrary metric linear spaces (Theorem 4.2 below) in contrast to the situation encountered in the classical Uniform Boundedness Principle.

<u>Definition 1</u>. Let (E,τ) be a topological group. A sequence $\{x_i\}$ in E is a τ- $\mathcal{K}$ convergent sequence if each subsequence of $\{x_i\}$ has a subsequence $\{x_{i_k}\}$ such that the series $\sum_k x_{i_k}$ is τ-convergent to an element $x \in E$.

If the topology τ is understood, we drop the τ in the description of τ- $\mathcal{K}$ convergence.

Note that any τ- K convergent sequence $\{x_i\}$ is τ-convergent to 0 by the <u>Urysohn property</u>, i.e., any subsequence of $\{x_i\}$ has a subsequence which converges to 0. In complete spaces the converse holds.

Notice in the Basic Matrix Theorem 2.2, assumption (II) implies that the rows of the matrix are K convergent (in some uniform sense). This is the reason for the terminology: K <u>matrix</u>.

<u>Example 2</u>. Let E be a complete normed group and $\{x_i\}$ converge to 0 in E. Then any subsequence of $\{x_i\}$ has a subsequence $\{x_{i_k}\}$ such that $\sum_k |x_{i_k}| < \infty$. The completeness implies that the series $\sum_k x_{i_k}$ converges in E. Thus, in complete spaces a sequence is K convergent iff it converges to 0.

In general the statement in Example 2 is false as the following example shows.

<u>Example 3</u>. Let c_{oo} be the vector space of all real sequences $\{t_j\}$ such that $t_j = 0$ eventually. Equip c_{oo} with the sup-norm. Let e_k be the sequence in c_{oo} which has a 1 in the kth coordinate and 0 elsewhere. Consider the sequence $\{(1/j)e_j\}$ in c_{oo}. This sequence converges to 0 in c_{oo} but no subseries of the series $\sum(1/j)e_j$ converges to an element of c_{oo}. That is, this sequence converges to 0 but is not K convergent.

Examples 2 and 3 might suggest that a (normed) space is complete iff it has the property that every sequence which converges to 0 is K convergent. There are, however, normed spaces which have this property but are not complete. A topological group which has the property that any sequence which converges to 0 is K convergent is called a K <u>space</u>. Klis ([45]) has given an example of a normed

space which is a $\mathcal{K}$ space but is not complete. (See also [26],
Theorem 2 and [48] Theorem 1.) The example given by Klis is fairly
involved and we do not present it here since it is not needed in the
sequel.

The notion of a $\mathcal{K}$ space was originally introduced in another
equivalent form by S. Mazur and W. Orlicz in [52], Axiom II, p. 169,
where it was observed that the classical Uniform Boundedness Principle holds for such spaces. A. Alexiewicz also studied this notion in
[1], Axiom $A_2^{\backprime}$, p. 203.

We record here several other interesting properties of $\mathcal{K}$
spaces and give references to the literature where these results can
be found. We do not present the results here since they are not
germane to the theme of these lecture notes.

First, any metric group which is a $\mathcal{K}$ space is also a Baire
space ([26] Theorem 2). More generally, any Frechet topological
group is also a Baire space ([26] Theorem 2). Second, there exist
Baire spaces which are not $\mathcal{K}$ spaces. Indeed, Burzyk, Klis and
Lipecki have shown that any infinite dimensional F-space contains a
subspace which is a Baire space but is not a $\mathcal{K}$ space ([26] Theorem
3).

Recall that a subset B of a topological vector space is
bounded iff for each sequence $\{x_j\} \subseteq B$ and each sequence of scalars
$\{t_j\}$ which converges to 0, the sequence $\{t_j x_j\}$ converges to 0.
Using this criteria and the notion of $\mathcal{K}$ convergence, we can introduce the concept of a $\mathcal{K}$ bounded set.

Definition 4. Let (E,τ) be a topological vector space. A subset
$B \subseteq E$ is τ- $\mathcal{K}$ bounded iff for each sequence $\{x_j\} \subseteq B$ and each
sequence of real numbers $\{t_j\}$ such that $\lim t_j = 0$, the sequence
$\{t_j x_j\}$ is τ- $\mathcal{K}$ convergent.

Since a $\mathcal{K}$ convergent sequence converges to 0, any $\mathcal{K}$ bounded set is always bounded. The converse is false; consider the sequence $\{e_j\}$ in Example 3. This sequence is bounded in c_{oo} but is not $\mathcal{K}$ bounded (by the argument in Example 3). By the observation in Example 2, if follows that in an F-space a subset is bounded iff it is $\mathcal{K}$ bounded.

As will be noted later the notion of $\mathcal{K}$ boundedness is quite useful, particularly in reformulating the classical uniform boundedness principle (Corollary 4.4). There are, however, some annoying difficulties associated with the notion of $\mathcal{K}$ boundedness. A convergent sequence in a topological vector space is always bounded; however, a $\mathcal{K}$ convergent sequence needn't be $\mathcal{K}$ bounded. We present an example of this phenomena in the example below.

Example 5. Let m_o be the space of all real sequences $\{t_j\}$ such that $\{t_j : j \in \mathbb{N}\}$ is finite. Pick $\{\phi_k\} \in \ell^1$ with $\phi_k \neq 0$ for each k. Define a norm (induced by $\{\phi_k\}$) on m_o by $\|\{t_j\}\| = \sum_j |\phi_j t_j|$. Consider the sequence $\{e_j\}$ in m_o. The series Σe_j is $\| \ \|$-subseries convergent in m_o (because

$$\|\sum_{j=1}^{n} e_{k_j} - C_{\{k_j : j \in \mathbb{N}\}}\| = \sum_{j=n+1}^{\infty} |\phi_{k_j}| \to 0$$

for each subsequence, where C_E denotes the characteristic function of E), and, therefore, $\{e_j\}$ is $\| \ \|$- $\mathcal{K}$ convergent in m_o. If $\{s_j\} \in c_o$ and the $\{s_j : j \in \mathbb{N}\}$ are distinct, then no subseries $\sum_{j=1}^{\infty} s_{k_j} e_{k_j}$ is $\| \ \|$-convergent to an element of m_o (indeed $\sum_{j=1}^{\infty} s_k e_k$ converges coordinatewise to an element of $\ell^\infty \setminus m_o$).

Hence, $\{e_j\}$ is $\| \ \|$- $\mathcal{K}$ convergent but not $\| \ \|$- $\mathcal{K}$ bounded.

The construction in Example 5 can be generalized to give a large number of examples of $\mathcal{K}$ convergent sequences which are not $\mathcal{K}$ bounded. Let Σ be a σ-algebra of subsets of a set S. Let $v : \Sigma \to \mathbf{R}$ be a positive, bounded, finitely additive set function. If $S(\Sigma)$ is the space of all Σ-simple functions, then v induces a semi-norm on $S(\Sigma)$ by $||\emptyset||_v = ||\emptyset|| = \int |\emptyset| dv$. Let $\{E_j\}$ be any disjoint sequence from Σ and consider the sequence $\{C_{E_j}\}$ in $S(\Sigma)$, where C_E denotes the characteristic function of E. By Drewnowski's Lemma 1.1, any subsequence of $\{E_j\}$ has a subsequence $\{E_{j_k}\}$ such that v is countably additive on the σ-algebra generated by $\{E_{j_k}\}$.

Thus, $\{C_{E_j}\}$ is $||\ ||$- $\mathcal{K}$ convergent (since $|| \sum_{k=1}^{n} C_{E_{j_k}} - C_{UE_{j_k}} || \to 0$ by countable additivity). However, if $\{t_j\} \in c_o$ and the $\{t_j : j \in \mathbf{N}\}$ are distinct, then no subseries of $\sum_j t_j C_{E_j}$ is $||\ ||$- $\mathcal{K}$ convergent to an element of $S(\Sigma)$. Thus, $\{C_{E_j}\}$ is $||\ ||$- $\mathcal{K}$ convergent but not $||\ ||$- $\mathcal{K}$ bounded.

This construction actually shows that the Drewnowski Lemma can be viewed as a result concerning $\mathcal{K}$ convergence. Namely, if $\{E_j\}$ is a disjoint sequence from Σ, then the sequence $\{C_{E_j}\}$ is $||\ ||_v$- $\mathcal{K}$ convergent for any v. More generally, if v_i is a sequence of positive, bounded, finitely additive measures on Σ and τ is the metric topology on $S(\Sigma)$ induced by the sequence of semi-norms $\{||\ ||_{v_i}\}$, then $\{C_{E_j}\}$ is τ- $\mathcal{K}$ convergent.

In the remainder of this section we collect some miscellaneous results concerning $\mathcal{K}$ convergence which, hopefully, will illustrate some of the applications of the notion of $\mathcal{K}$ convergence.

First, it is well known that in the dual of a normed space X, a set may be weak* bounded but not norm bounded. For example, in the dual, ℓ^1, of c_{oo}, the sequence $\{je_j\}$ is weak* bounded but not norm bounded. However, for weak*- $\mathcal{K}$ bounded sets we have the following result.

Theorem 6. Let X be a normed space. If $B \subseteq X'$ is weak*- $\mathcal{K}$ bounded, then B is norm bounded (and, hence, norm- $\mathcal{K}$ bounded since X' is complete).

Proof: Let $\{x_j'\} \subseteq B$ and $\{t_j\}$ be a sequence of scalars which converges to 0. For each j pick $x_j \in X$ such that $||x_j|| = 1$ and $|<x_j', x_j>| \geqslant ||x_j'|| - 1/j$. Consider the matrix $[z_{ij}] = [<\sqrt{|t_j|}x_j', \sqrt{|t_i|}x_i>]$. Since $\sqrt{|t_i|}x_i \to 0$ in norm and $(\sqrt{|t_j|}x_j')$ is weak*- $\mathcal{K}$ convergent, the matrix $[z_{ij}]$ is a $\mathcal{K}$ matrix. By the Basic Matrix Theorem,
$$\lim_i z_{ii} = \lim_i |t_i|<x_i', x_i> = 0 \quad \text{so that} \quad \lim_i |t_i| \, ||x_i'|| = 0.$$

Thus, B is norm bounded.

The proof above also shows that the (undesirable) phenomenon displayed in Example 5 does not occur in weak* topologies. That is, the proof above shows that a weak*- $\mathcal{K}$ convergent sequence $\{x_j'\}$ is norm bounded and since X' is complete, the sequence $\{x_j'\}$ is also norm- $\mathcal{K}$ bounded and, therefore, weak*- $\mathcal{K}$ bounded.

Another interesting property of $\mathcal{K}$ convergence concerns the weak topology in a normed space. Again it is well known that in infinite dimensional normed spaces a sequence which converges to 0 weakly will generally not be norm convergent. (ℓ^1 is an interesting exception to this statement.) For example, the sequence $\{e_j\}$ in c_o is weakly convergent to 0 but is certainly not norm convergent. For weak- $\mathcal{K}$ convergent sequences, we, however, have

Theorem 7. Let X be a normed space. If $\{x_j\} \subseteq X$ is weak- $\mathcal{K}$
convergent, then $\{x_j\}$ converges to 0 in norm.

Proof: By replacing X by the closed linear subspace generated by
the $\{x_j\}$, we may assume that X is separable. For each j pick
$x_j' \in X'$ such that $||x_j'|| = 1$ and $\langle x_j', x_j \rangle = ||x_j||$. By the Banach-
Alaogu Theorem, $\{x_j'\}$ has a subsequence $\{x_{k_j}'\}$ which converges weak*
to an element $x' \in X'$. Consider the matrix $[z_{ij}] = [\langle x_{k_i}', x_{k_j} \rangle]$.
By the weak* convergence of $\{x_{k_j}'\}$ and the weak- $\mathcal{K}$ convergence of
$\{x_j\}$, $[z_{ij}]$ is a $\mathcal{K}$ matrix. By the Basic Matrix Theorem,
$\lim z_{ii} = \lim ||x_{k_i}|| = 0$. Since the same argument can be applied
to any subsequence of $\{x_i\}$, the Urysohn property implies that
$\lim ||x_i|| = 0$.

Actually, Theorem 7 can be improved to conclude that a sequence
in a normed space is weak- $\mathcal{K}$ convergent iff it is norm- $\mathcal{K}$ convergent.
That is, we have

Corollary 8. Let X be a normed space. Then a sequence in X is
weak- $\mathcal{K}$ convergent iff it is norm- $\mathcal{K}$ convergent.

Proof: Clearly norm- $\mathcal{K}$ convergence implies weak- $\mathcal{K}$ convergence.

Suppose $\{x_n\}$ is weak- $\mathcal{K}$ convergent. By Theorem 7, $\{x_n\}$ is
norm convergent to 0. Let $\{y_n\}$ be a subsequence of $\{x_n\}$ such
that $||y_n|| \leqslant 1/2^n$. Next, let $\{z_n\}$ be a subsequence of $\{y_n\}$ and
$z \in X$ be such that

(1) $$\sum_{n=1}^{\infty} z_n = z,$$

where the convergence is in the weak topology. But,

$$\sum_{i=m}^{n} ||z_i|| \leqslant \sum_{i=m}^{n} 1/2^i$$

so the series Σz_i is norm Cauchy. Consequently, (1) holds with respect to the norm topology.

We indicate an application of Theorem 7 by using Theorem 7 to derive the classical Orlicz-Pettis Theorem. Recall that a series Σx_i in a normed space X is norm (weakly) <u>subseries</u> <u>convergent</u> if each subseries Σx_{k_i} is norm (weakly) convergent in X. The remarkable Orlicz-Pettis Theorem ([55], [59]) is given by

<u>Theorem 9</u>. Let X be a normed space. If Σx_i is weak subseries convergent in X, then Σx_i is norm subseries convergent.

<u>Proof</u>: It suffices to show that $||x_i|| \to 0$ (See Theorem 7.1 or [30] IV. 1.1; Orlicz-Pettis results are discussed at some length in length in section 7). But $\{x_i\}$ is clearly weak-$\mathcal{K}$ convergent so Theorem 7 gives the result.

Note that the analogue of Theorem 7 is false for the weak* topology. For example, the sequence $\{e_j\}$ is weak*-$\mathcal{K}$ convergent in ℓ^∞ but is certainly not norm convergent. Concerning the weak* topology, we do, however, have

<u>Theorem 10</u>. Suppose X is a normed space which contains no subspace isomorphic (topologically) to ℓ^1. If $\{x_j'\} \subseteq X'$ is weak*-$\mathcal{K}$ convergent, then $\{x_j'\}$ converges to 0 in norm.

<u>Proof</u>: Pick $x_j \in X$ such that $||x_j|| = 1$ and $<x_j',x_j> + (1/j) > ||x_j'||$. Since X contains no subspace isomorphic to ℓ^1, the se-

quence $\{x_j\}$ contains a subsequence $\{x_{k_j}\}$ which is weak Cauchy

([49] 2.e.5). Consider the matrix $z_{ij} = \langle x_{k_j}', x_{k_i}' \rangle$. From the facts

that $\{x_{k_j}\}$ is weak Cauchy and $\{x_j'\}$ is weak*- $\mathcal{K}$ convergent, it

follows that $[z_{ij}]$ is a $\mathcal{K}$ matrix. The Basic Matrix Theorem im-

plies that $\lim_i z_{ii} = 0$ so that $\lim ||x_{k_i}'|| = 0$. Since the same

argument can be applied to any subsequence of $\{x_i'\}$, the Urysohn

property implies that $\lim ||x_i'|| = 0$.

This result is somewhat analagous to the Diestel-Faires Theorem

on weak* subseries convergence ([33] or Theorem 10.10).

It is not known if the converse of Theorem 10 holds.

We conclude this section by establishing a result due to E. Pap

concerning the boundedness of adjoint operators (private communica-

tion). Recall that if X and Y are normed spaces and T is a

linear operator with domain, D(T), a dense subspace of X and range

in Y, then the adjoint operator T' is defined by the following:

the domain, $D(T')$, of T' is $D(T') = \{y' \in Y' : y'T$ is

continuous on $D(T)\}$ and $T' : D(T') \to X'$ is defined by $T'y'$ is

the unique continuous extension of $y'T$ to X. (It is necessary

that D(T) be dense in X for this definition to make sense.)

Thus, for $x \in D(T)$, $y' \in D(T')$, we have $\langle T'y',x \rangle = \langle y',Tx \rangle$; this

formula agrees with the usual definition of the adjoint of a bounded

linear operator. It can happen, when T is not continuous, that

$D(T')$ consists of only the 0 vector; however, if T is a closed

operator, then $D(T')$ is weak* dense in Y'. Concerning the adjoint

operator, we have the following interesting result of E. Pap.

Theorem 11. Let X be a $\mathcal{K}$ - space and $T : X \to Y$ be linear.

Then T' is a bounded linear operator on $D(T')$.

<u>Proof</u>: It suffices to show that if $\{y_i'\} \subseteq D(T')$ and $||y_i'|| \leqslant 1$, then $\{T'y_i'\}$ is bounded. Pick $x_i \in X$ such that $||x_i|| = 1$ and $||T'y_i'|| \leqslant <T'y_i', x_i> + 1$. Thus, it suffices to show that $\{<T'y_i', x_i>\}$ is bounded. For this, it suffices to show that if $\{t_i\}$ is a positive sequence of scalars which converges to 0, then $\{t_i <T'y_i', x_i>\}$ converges to 0.

Put $z_{ij} = <\sqrt{t_i} T'y_i', \sqrt{t_j} x_j> (= <\sqrt{t_i}y_i', T(\sqrt{t_j}x_j)>)$. We show that $\{z_{ij}\}$ is a $\mathcal{K}$ - matrix. First for each i,

$\lim\limits_j z_{ij} = \lim\limits_j <\sqrt{t_i}T'y_i', \sqrt{t_j}x_j> = 0$ since $\sqrt{t_j} x_j \to 0$. Next, for each

j, $\lim\limits_i z_{ij} = \lim\limits_i <\sqrt{t_i}y_i', T(\sqrt{t_j}x_j)> = 0$ since $\sqrt{t_i}y_i' \to 0$. Now

$\sqrt{t_i}x_j \to 0$ and X is a $\mathcal{K}$ - space so there is a subsequence $\{k_j\}$ such that the subseries $\sum\limits_j \sqrt{t_{k_j}}x_{k_j}$ converges in norm to an element

$x \in X$. Then,

$$\sum_j z_{ik_j} = \sum_j <\sqrt{t_i}\, T'y_i'\, \sqrt{t_{k_j}}x_{k_j}>$$

$$= <\sqrt{t_i}T'y_i', x>$$

$$= <\sqrt{t_i}y_i', Tx> \to 0$$

since $\sqrt{t_i}y_i' \to 0$. Thus, $\{z_{ij}\}$ is a $\mathcal{K}$-matrix.

4. The Uniform Boundedness Principle

In this section we discuss the well-known Uniform Boundedness Principle (UBP). We give a proof of the UBP based on the Basic Matrix Theorem (2.2) instead of the familiar Baire Category proof which is so often given (cf. for example, [19], [38] II.1.11). By employing matrix methods we are actually able to establish a version of the UBP which is valid with no completeness assumptions and which yields the classical UBP for F-spaces as an immediate corollary. In order to indicate how the general UBP can be used in the absence of completeness, we derive a version of the Nikodym Boundedness Theorem of measure theory from the general UBP.

We first state the classical UBP for normed spaces in order to motivate the discussion of the general UBP.

__Theorem 1__. Let X be a B-space. If $\{T_i\} \subseteq L(X,Y)$ is pointwise bounded on X, then $\{T_i\}$ is uniformly bounded on the $\| \; \|$-bounded subsets of X.

The conclusion of the norm version of the UBP is usually written in the form : $\{\|T_i\|\}$ is bounded, where $\|T\|$ is the operator norm of T. This conclusion clearly implies the conclusion in Theorem 1 since if $B \subseteq X$ is bounded, then

$$\|T_i x\| \leqslant \sup_i \|T_i\| \sup_B \|x\|$$

for each i and $x \in B$.

It is well-known that the conclusion in Theorem 1 is false if the completeness assumption on X is dropped. For example, the sequence of linear functionals $\{f_i\}$ defined on c_{oo} by $< f_i, \{t_j\} >$ $= it_i$ is pointwise bounded but not uniformly bounded on the bounded subset $\{e_j : j \in \mathbb{N}\}$ of c_{oo}.

Thus, if one is to seek a version of the UBP which is valid for arbitrary normed spaces X, the family of bounded subsets of X is too large to insure that a pointwise bounded sequence of operators is uniformly bounded on each member of the family. It would be desirable to have a family of subsets of X with the property that a pointwise bounded sequence of operators is uniformly bounded on each member of the family. In order to be interesting this family should be as large as possible, and it would be of interest to find a family of sets which coincides with the family of bounded sets when the space X is complete. It is shown in Corollary 4 that the family of $\mathcal{K}$ bounded sets has this property.

We first establish our general UBP. If $\Gamma \subseteq L(X,Y)$, let $\tau(\Gamma)$ be the weakest topology on X such that each member of Γ is continuous; thus, a sequence $\{x_j\}$ in X converges to 0 in $\tau(\Gamma)$ if and only if $Tx_j \to 0$ for each $T \in \Gamma$. In order to shorten the notation we write $\Gamma-\mathcal{K}$ bounded ($\Gamma-\mathcal{K}$ convergent) for $\tau(\Gamma)-\mathcal{K}$ bounded ($\tau(\Gamma)-\mathcal{K}$ convergent).

Theorem 2. Let X,Y be metric linear spaces and let $T_i \in L(X,Y)$ be such that $\{T_i x\}$ is bounded for each $x \in X$. Then

(i) $\{T_i\}$ is uniformly bounded on $\{T_i\}-\mathcal{K}$ convergent sequences and

(ii) $\{T_i\}$ is uniformly bounded on $\{T_i\}-\mathcal{K}$ bounded sets.

Proof: If (i) is false, there is a balanced neighborhood U of 0 in Y and a $\{T_i\}-\mathcal{K}$ convergent sequence $\{x_j\}$ such that $\{T_i x_j\}$ is not absorbed by U. Thus, there exists positive integers m_1 and n_1 such that $T_{m_1} x_{n_1} \notin U$. Set $k_1 = 1$. Then by the pointwise boundedness and the fact that $\lim_j T_i x_j = 0$ for each i, there exists

$k_2 > k_1$ such that $\{T_i x_j : 1 \leqslant i \leqslant m_1, \ 1 \leqslant j < \infty\} \cup \{T_i x_j : 1 \leqslant i < \infty, \ 1 \leqslant j \leqslant n_1\} \subseteq k_2 U$. By assumption, there exist m_2 and n_2 such that $T_{m_2} x_{n_2} \notin k_2 U$. Note that $m_2 > m_1$ and $n_2 > n_1$. Continuing this construction produces subsequences $\{m_i\}$, $\{n_i\}$ and $\{k_i\}$ such that $T_{m_i} x_{n_i} \notin k_i U$. If $t_i = 1/k_i$, then $\{t_i\} \in c_0$.

Now consider the matrix $[t_i T_{m_i} x_{n_j}]$. By the pointwise bounded-ness assumption and the $\{T_i\}$- $\mathcal{K}$ convergence of the sequence $\{x_j\}$, this is a $\mathcal{K}$ matrix. Hence, by Theorem 2.2, $\lim_i t_i T_{m_i} x_{n_i} = 0$.

Thus, $t_i T_{m_i} x_{n_i} \in U$ for large i. This contradicts the construction and establishes (i).

For (ii), let B be $\{T_i\}$- $\mathcal{K}$ bounded. To show $\{T_i B\}$ is bounded, as in part (i), it suffices to show $\{T_i x_i\}$ is bounded for each $\{x_i\} \subseteq B$. Let $\{t_i\} \in c_0$. Then $\{\sqrt{|t_i|} \, x_i\}$ is $\{T_i\}$- $\mathcal{K}$ convergent so part (i) implies $\{T_i(\sqrt{|t_i|} \, x_i)\}$ is bounded. Hence, $\lim_i \sqrt{|t_i|} \, T_i(\sqrt{|t_i|} \, x_i) = \lim_i |t_i| T_i x_i = 0$ so that $\{T_i x_i\}$ is bounded.

Note that the conclusion in Theorem 2 is much sharper than the conclusion of the classical UBP in the sense that the family of sub-sets of X where the given sequence $\{T_i\}$ is uniformly bounded depends very much on the particular sequence $\{T_i\}$. Indeed, given a sequence $\{T_i\}$, the family of $\{T_i\}$- $\mathcal{K}$ convergent sequences ($\{T_i\}$- $\mathcal{K}$ bounded sets) may even contain subsets of X which are not norm-bounded. (It is easy to check that if a family of subsets of a normed space X has the property that <u>any</u> pointwise bounded sequence of continuous linear functionals on X is uniformly bounded on each member of the family, then each member of the family is a bounded set. Indeed, if $\{x_i\}$ is an unbounded sequence in a normed space

X, pick $x_i' \in X'$ such that $\langle x_i', x_i \rangle = ||x_i||$ and $||x_i'|| = 1$. Then $\{x_i'\}$ is uniformly bounded on bounded subsets of X but is not uniformly bounded on the sequence $\{x_i\}$.) For example, the sequence $\{i\,e_{2i}\}$ in $\ell^1 = (c_{oo})'$ is pointwise bounded on c_{oo} and is also uniformly bounded on the unbounded subset $\{j\,e_{2j+1} : j \in \mathbb{N}\}$ of c_{oo}.

In order to obtain the largest possible family $\mathcal{J}$ with the property that each pointwise bounded sequence of operators is uniformly bounded on each member of $\mathcal{J}$, we use the $\tau(L(X,Y))$ topology of X.

Corollary 3. Let $\{T_i\} \subseteq L(X,Y)$ be pointwise bounded. Then

(i) $\{T_i\}$ is uniformly bounded on $L(X,Y)$- $\mathcal{K}$ convergent sequences and

(ii) $\{T_i\}$ is uniformly bounded on $L(X,Y)$- $\mathcal{K}$ bounded sets.

It is worthwhile noting that the families of $\mathcal{K}$ convergent sequences ($\mathcal{K}$ bounded sets) in Theorem 2 and Corollary 3 are in general not the same. That is, for certain topologies these families can be different. For example, in m_o the sequence $\{e_i\}$ is $\{\nu_i\}$- $\mathcal{K}$ convergent for any given sequence $\{\nu_i\}$ in $ba = m_o$ by Drewnowski's Lemma. However, $\{e_i\}$ is not ba- $\mathcal{K}$ convergent since given any subsequence $\{n_k\}$ of positive integers, there exists a finitely additive measure μ on the power set $\mathcal{P}$ of $\mathbb{N}$ such that $\mu(\bigcup_k\{n_k\}) = 1$ and $\mu(\{n_k\}) = 0$ for each k. (The existence of such a measure μ can be shown as follows: Let f be the one-one map $k \rightarrow n_k$ from $\mathbb{N}$ onto $\{n_k : k \in \mathbb{N}\} = R$. Let σ be a $0 - 1$ finitely additive measure on $\mathcal{P}$ ([41] p. 358, 20. 38). Define μ by $\mu(A) = \sigma(f^{-1}(A))$. Then $\mu(R) = 1$ but $\mu(\{n_k\}) = 0$ for each k.) Thus, the family of $\{T_i\}$- $\mathcal{K}$ convergent sequences in Theorem 2 is, in general, larger than the family of $L(X,Y)$- $\mathcal{K}$ convergent sequence in Corollary 3.

Since the $\tau(L(X,Y))$ topology of X is weaker than the natural topology of X, we obtain immediately.

Corollary 4. Let $\{T_i\} \subseteq L(X,Y)$ be pointwise bounded. Then

(i) $\{T_i\}$ is uniformly bounded on $||$- K convergent sequences

and

(ii) $\{T_i\}$ is uniformly bounded on $||$- K bounded sets.

Since the families of norm bounded sets and K bounded sets co-incide in B-spaces, Corollary 4 in particular contains the UBP for B-spaces given in Theorem 1 as a Corollary. More generally, Corollary 4 yields the UBP for K spaces as presented by Mazur and Orlicz ([52], p. 169) (recall that a metric linear space is a K space if every sequence which converges to 0 is a K sequence). Again it should be noted that the notions of K convergent sequence and K bounded set allow one to formulate versions of the UBP which are valid in arbitrary metric linear spaces. This should be con-trasted with the classical UBP as well as the version of the UBP given by Mazur and Orlicz in [52].

Corollary 4 now yields immediately the classical UBP for F-spaces.

Corollary 5. Let X be an F-space and let $\{T_i\} \subseteq L(X,Y)$ be point-wise bounded. Then

(i) $\{T_i\}$ is uniformly bounded on $||$-bounded subsets of X and

(ii) $\{T_i\}$ is equicontinuous.

Proof: Conclusion (i) follows from Corollary 4 (ii) since in a complete space the families of bounded sets and K bounded sets coincide.

To establish (ii), it suffices to show that $T_i x_i \to 0$ whenever

$x_i \to 0$ in X. Let $x_i \to 0$ in X and pick a sequence of positive scalars $\{t_i\}$ such that $t_i \to \infty$ and $t_i x_i \to 0$. By (i), $\{T_i(t_i x_i)\}$ is bounded. Consequently, $(1/t_i)\ T_i\ (t_i x_i) = T_i x_i \to 0$ since $\{1/t_i\} \in c_o$.

The conclusion in (ii) is the conclusion in the usual statement of the classical UBP for F-spaces (cf. [38] II.1.11).

To indicate an application of the general UBP in Theorem 2, we derive a version of the Nikodym Boundedness Theorem. The classical Nikodym Boundedness Theorem is given in the following statement:

Theorem 6. Let Σ be a σ-algebra of subsets of a set S and let $\mu_i : \Sigma \to \mathbf{R}$ be countably additive. If $\{\mu_i(E)\}$ is bounded for each $E \in \Sigma$, then $\{\mu_i(E) : i \in \mathbf{N},\ E \in \Sigma\}$ is bounded.

This result is called "a striking improvement of the UBP" for the space $ca(\Sigma)$ by Dunford and Schwartz ([38] III.1.5). We recast this theorem in a form which is more analogous to the usual statement in the UBP and then indicate why the result does not follow from the classical UBP.

Let $S(\Sigma)$ be the space of all Σ-simple functions equipped with the sup-norm. Each μ_i induces a continuous linear functional f_i on $S(\Sigma)$ via integration, $\langle f_i, y \rangle = \int y d\mu_i$. The dual norm of f is the variation of μ_i, $|\mu_i|$, on S, i.e., $||f_i|| = |\mu_i|(S)$. This variation norm is, however, equivalent to the norm $||\mu_i||' = \sup\{|\mu_i(E)| : E \in \Sigma\}$. In this notation, Theorem 6 takes the familiar form: if $\{f_i\}$ is pointwise bounded on $S(\Sigma)$, then $\{||f_i||\}$ is bounded.

Note that this result is not obtainable from Theorem 1 since the space $S(\Sigma)$ is not complete (or even 2nd category [66]). We will derive Theorem 6 from the general UBP given in Theorem 2.

Actually, Darst ([34]) has shown that Theorem 6 is even valid for bounded, finitely additive set functions, and, furthermore, it is known that Theorem 6 is also valid for certain algebras of sets which are not σ-algebras ([67]; see also Theorem 5.12). We will derive a version of the Nikodym Boundedness Theorem for bounded finitely additive set functions on σ-algebras.

For the proof of the Nikodym Boundedness Theorem, we require the following "well-known" lemma ([37]).

Lemma 7. Let A be an algebra of subsets of a set S and let $\mu_i : A \rightarrow \mathbb{R}$ be bounded and finitely additive. Then $\{\mu_i(E) : i \in \mathbb{N}, E \in A\}$ is bounded if and only if $\{\mu_i(E_j) : i, j \in \mathbb{N}\}$ is bounded for each disjoint sequence $\{E_i\} \subseteq A$.

Proof: Suppose $\sup \{ |\mu_i(E)| : i \in \mathbb{N}, E \in A\} = \infty$. Note that for each $M > 0$ there is a partition (E,F) of S and an integer i such that $\min \{ |\mu_i(E)|, |\mu_i(F)| \} > M$. (This follows since $|\mu_i(E)| > M + \sup \{ |\mu_i(S)| : i \in \mathbb{N}\}$ implies $|\mu_i(S \setminus E)| \geqslant |\mu_i(E)| - |\mu_i(S)| > M$.) Hence, there exist i_1 and a partition (E_1, F_1) of S such that $\min\{ |\mu_{i_1}(E_1)|, |\mu_{i_1}(F_1)| \} > 1$. Now either $\sup \{ |\mu_i(H \cap E_1)| : H \in A, i \in \mathbb{N}\} = \infty$ or $\sup \{ |\mu_i(H \cap F_1)| : H \in A, i \in \mathbb{N}\} = \infty$. Pick whichever of E_1 or F_1 satisfies this condition and label it B_1 and set $A_1 = S \setminus B_1$. Now treat B_1 as S above to obtain a partition (A_2, B_2) of B_1 and an $i_2 > i_1$ satisfying $|\mu_{i_2}(A_2)| > 2$ and $\sup \{ |\mu_i(H \cap B_2)| : H \in A, i \in \mathbb{N}\} = \infty$. Proceeding by induction produces a subsequence $\{i_j\}$ and a disjoint sequence $\{A_j\}$ such that $|\mu_{i_j}(A_j)| > j$.

This establishes the sufficiency; the necessity is clear.

The Nikodym Boundedness Theorem for finitely additive set functions is given by

Theorem 8. Let Σ be a σ-algebra and let $\mu_i : \Sigma \to \mathbb{R}$ be bounded and finitely additive. If $\{\mu_i(E)\}$ is bounded for each $E \in \Sigma$, then $\{\mu_i(E) : i \in \mathbb{N}, E \in \Sigma\}$ is bounded.

Proof: By Lemma 7, it suffices to show that $\{\mu_i(E_j) : i, j \in \mathbb{N}\}$ is bounded for each disjoint sequence $\{E_j\} \subseteq \Sigma$. Let $\{E_j\}$ be a disjoint sequence from Σ.

Let $S(\Sigma)$ be the space of Σ-simple functions equipped with the sup-norm. Each μ_i induces a continuous linear functional f_i on $S(\Sigma)$ via integration, $\langle f_i, \phi \rangle = \int \phi d\mu_i$. Consider the sequence $\{C_{E_j}\}$ in $S(\Sigma)$, where C_E denotes the characteristic function of E. By Drewnowski's Lemma 1.1, any subsequence of $\{E_j\}$ has a subsequence $\{F_j\}$ such that each $\{\mu_i\}$ is countably additive on the σ-algebra generated by $\{F_j\}$. That is, $\{C_{E_j}\}$ is $\{f_i\}$- $\mathcal{K}$ convergent in $S(\Sigma)$. By Theorem 2 (i), $\{f_i(C_{E_j})\} = \{\mu_i(E_j)\}$ is bounded, and the result is established.

Theorem 8 is also valid for measures with values in locally convex spaces, and the vector version of Theorem 8 can easily be obtained from Theorem 8 by employing the classical UBP ([34] I.3.1).

The Nikodym Boundedness Theorem is also valid for measures defined on certain algebras that are not necessarily σ-algebras. For example, see [67] and the remarks following Theorem 5.10.

In conclusion, it should also be noted that Corollary 4 cannot be used in the proof of Theorem 8 given above since the sequence $\{C_{E_j}\}$ is clearly not $\|\ \|$- $\mathcal{K}$ convergent.

5. Convergence of Operators

In this section we consider two classical results concerning
convergent sequences of operators which are usually attributed to
Banach and Steinhaus. We first note that one of the results, usually
referred to as the Banach-Steinhaus Theorem, follows immediately from
the results on the UBP in section 4.

Theorem 1. (Banach-Steinhaus) Let X be an F-space and let
$\{T_i\} \subseteq L(X,Y)$. If $\lim\limits_i T_i x = Tx$ exists for each x, then $T : X \to Y$
is continuous and linear.

Proof: From Corollary 4.5, the sequence $\{T_i\}$ is equicontinuous.
Let $\{x_j\}$ converge to 0 in X. Then, from the equicontinuity, we
have $\lim\limits_j Tx_j = \lim\limits_j \lim\limits_i T_i x_j = \lim\limits_i \lim\limits_j T_i x_j = 0$. Hence, T is
continuous.

We next consider another result which is sometimes attributed to
Banach and Steinhaus ([47] 39.5). This result also pertains to convergent sequences of operators and is usually presented under either
barrelledness or completeness assumptions on the domain space. We
present a version of the result which is valid without any completeness assumptions; the proof is again based on the Basic Matrix
Theorem instead of the familiar Baire category methods. Our general
result yields the classical result of Banach-Steinhaus for F-spaces
as an immediate corollary. To indicate an application of our general
result, we use it to derive a version of the Nikodym Convergence
Theorem and also consider two classical results in summability due to
Hahn and Schur.

We begin by giving a statement of the classical result for

F-spaces. The result is sometimes given for barrelled spaces as well ([47] 39.5).

Theorem 2. (Banach-Steinhaus) Let X be an F-space. If $\{T_i\}$ $\subseteq L(X,Y)$ is such that $\lim_i T_i x = Tx$ exists for each $x \in X$, then

T is continuous and $\lim_i T_i x = Tx$ converges uniformly for x

belonging to any compact subset of X.

Again it is well-known that this result is false if the completeness assumption is dropped. For example, the sequence $\{i^2 e_i\}$ in $\ell' = (c_{oo})'$ converges pointwise to 0 on c_{oo} but the convergence is not uniform on the relatively compact sequence $\{(1/j)e_j\} \subseteq$ c_{oo}. Using the notion of $\mathcal{K}$ convergence, we formulate a version of Theorem 2 which is valid in the absence of completeness and which contains Theorem 2 as an immediate corollary. We will refer to Theorem 3 below as the General Banach-Steinhaus Theorem.

Theorem 3. (General Banach-Steinhaus Theorem) Let $T_i \in L(X,Y)$. If $\lim_i T_i x = Tx$ exists for each $x \in X$, then $\lim_i T_i x = Tx$ uniformly for x belonging to any $\{T_i\}$- $\mathcal{K}$ convergent sequence $\{x_j\}$ in X. (T is not assumed to belong to $L(X,Y)$.)

Proof: Let $\{x_j\}$ be $\{T_i\}$- $\mathcal{K}$ convergent. Consider the matrix $[T_i x_j]$. By the pointwise convergence of $\{T_i\}$ and the $\{T_i\}$- $\mathcal{K}$ convergence of $\{x_j\}$, this is a $\mathcal{K}$ matrix. The Basic Matrix Theorem 2.2 implies $\lim_i T_i x_j = Tx_j$ uniformly in j and establishes the result.

One can also introduce the notion of " $\mathcal{K}$ compactness" or " $\mathcal{K}$ relative compactness" and give a formulation of Theorem 3 which is more analogous to the classical statement in Theorem 2. If (E,τ)

is a topological group, then a subset A of E is said to be $\tau - \mathcal{K}$ compact ($\tau - \mathcal{K}$ relatively compact) if every sequence in A has a subsequence $\{x_{n_i}\}$ which is $\tau - \mathcal{K}$ convergent to an element

$x \in A (x \in E)$. Using the notion of $\mathcal{K}$ compactness, the conclusion of Theorem 3 can be sharpened to state:

$\lim T_i x = Tx$ uniformly for x belonging to $\tau(\{T_i\}) - \mathcal{K}$ relatively compact subsets of X .

Using Theorem 3 we can establish Theorem 2 as an immediate corollary.

Proof of Theorem 2: It suffices to consider the case when the compact subset of X is a sequence $\{x_j\}$. From the compactness, it suffices to consider the case when the sequence $\{x_j\}$ converges to some $x \in X$. The sequence $\{x_j - x\}$ is $|| - \mathcal{K}$ convergent by the completeness of X so Theorem 3 implies that $\lim_i T_i(x_j - x) = T(x_j - x)$ uniformly in j. Since $\lim_i T_i x = Tx$, this means $\lim_i T_i x_j = Tx_j$ uniformly in j, and the proof of Theorem 2 is complete.

Since any sequence $\{x_j\} \subseteq X$ which is $|| - \mathcal{K}$ convergent in X is $\{T_i\} - \mathcal{K}$ convergent, Theorem 3 also has the following corollary which is perhaps more analogous to the statement in Theorem 2.

Corollary 4. Let $\{T_i\} \subseteq L(X,Y)$. If $\lim_i T_i x = Tx$ exists for each $x \in X$, then $\lim_i T_i x_j = Tx_j$ uniformly in j for any $|| - \mathcal{K}$ convergent sequence $\{x_j\}$ in X.

Remark 5. Notice that neither the scalar multiplication in X and Y or the homogeneity of the operators T_i was used in the statements or proofs above. Thus, the results above are actually valid for additive maps T_i between two normed groups X and Y. It is not known if the general versions of the results above for normed groups have any useful applications.

In order to illustrate the utility of Theorem 3 we give an application to a situation where completeness is not present. In particular, we consider a classical result of Nikodym on convergent sequences of measures ([38] III.7.4). This result of Nikodym, sometimes referred to as the Nikodym Convergence Theorem, is one of the most useful results in measure theory, and the original result for scalar measures has been generalized in several directions. For example, the result has been generalized to vector-valued measures and also to certain finitely additive set functions. We first consider one of these generalizations, sometimes referred to as the Brooks-Jewett Theorem ([24], [36]), to certain vector-valued finitely additive set functions.

Let Σ be an algebra of subsets of a set S and let Y be a locally convex metric linear space. A finitely additive set function $\mu : \Sigma \to Y$ is said to be <u>strongly</u> <u>additive</u> (<u>strongly bounded</u> or <u>exhaustive</u>) if $\lim \mu(E_j) = 0$ for each disjoint sequence $\{E_j\} \subseteq \Sigma$. A sequence $\{\mu_i\}$ of additive set functions is said to be <u>uniformly</u> <u>strongly additive</u> if $\lim_j \mu_i(E_j) = 0$ uniformly in i for each disjoint sequence $\{E_j\}$. A countably additive vector measure on a σ-algebra is clearly strongly additive, and a uniformly countably additive sequence of vector measures defined on a σ-algebra is also uniformly strongly additive. The Nikodym Convergence Theorem has been generalized to strongly additive set functions by Brooks and Jewett ([24], [36]); we first consider this generalization.

<u>Theorem 6</u>. Let Σ be a σ-algebra. Let $\mu_i : \Sigma \to Y$ be strongly

additive. If $\lim_i \mu_i(E) = \mu(E)$ exists for each $E \in \Sigma$, then

(i) μ is strongly additive and

(ii) $\{\mu_i\}$ is uniformly strongly additive.

Proof: Let $S(\Sigma)$ be the space of Σ-simple functions equipped with the sup-norm. Since each μ_i is bounded, each μ_i induces a continuous linear operator $T_i : S(\Sigma) \to Y$ by integration, $T_i \phi = \int_S \phi d\mu_i$.

(Note that even though Y is not assumed to be complete there is no elaborate integration theory required here since only simple functions are being integrated.) Let $\{E_j\}$ be a disjoint sequence from Σ. Consider the sequence $\{C_{E_j}\}$ in $S(\Sigma)$, where C_E denotes the characteristic function of E. By Drewnowski's Lemma 1.1, this sequence is $\{T_i\}$- $\mathcal{K}$ convergent. By Theorem 2,

$\lim_i T_i(C_{E_j}) = \lim_i \mu_i(E_j) = \mu(E_j)$ uniformly in j. Since

$\lim_j T_i(C_{E_j}) = \lim_j \mu_i(E_j) = 0$ for each i, we have $\lim_j \mu_i(E_j) = 0$

uniformly in i. This establishes (ii).

For (i), note that from the uniform convergence,

$\lim_j \mu(E_j) = \lim_j \lim_i \mu_i(E_j) = \lim_i \lim_j \mu_i(E_j) = 0$ so that μ is

strongly additive.

Remark 7. Note that Theorem 6 can be established directly from the Basic Matrix Theorem without referring to Theorem 2 by considering the matrix $[\mu_i(E_j)]$ and invoking the Drewnowski Lemma. The original version of the Brooks-Jewett Theorem was established for B-spaces whereas the version above does not assume completeness and is valid for locally convex metric linear spaces (see also [36]).

We now establish the usual version of the Nikodym Convergence

Theorem for vector-valued measures. A sequence $\{\mu_i\}$ of countably additive Y-valued measures, $\mu_i : \Sigma \to Y$, is said to be <u>uniformly</u> <u>countably</u> <u>additive</u> if $\lim_n \sum_{j=n}^{\infty} \mu_i(E_j) = 0$ uniformly in i for each disjoint sequence $\{E_j\}$. For the Nikodym Convergence Theorem, we require the following criteria for uniform countable additivity.

<u>Lemma 8.</u> Let Σ be a σ-algebra. Let $\mu_i : \Sigma \to Y$ be countably additive. The following are equivalent:

    (i) $\{\mu_i\}$ is uniformly countably additive

    (ii) for each decreasing sequence $\{E_j\}$ from Σ with

                $\cap E_j = \emptyset$, $\lim_j \mu_i(E_j) = 0$ uniformly in i.

    (iii) $\{\mu_i\}$ is uniformly strongly additive.

<u>Proof</u>: (i) and (ii) are clearly equivalent for countably additive measures and (i) clearly implies (iii).

    Suppose (iii) holds and (ii) fails to hold. Then we may assume (by passing to a subsequence if necessary) that there exist a decreasing sequence $\{F_j\}$ with $\cap F_j = \emptyset$ and a $\delta > 0$ such that $|\mu_i(F_i)| > \delta$ for each i. There exists k_1 such that $|\mu_1(F_{k_1})| < \delta/2$. Then there exists a $k_2 > k_1$ such that

$|\mu_{k_1}(F_{k_2})| < \delta/2$. Continuing by induction produces a subsequence

$\{k_j\}$ such that $|\mu_{k_j}(F_{k_{j+1}})| < \delta/2$. If $E_j = F_{k_j} \setminus F_{k_{j+1}}$,

then $\{E_j\}$ is a disjoint sequence from Σ with

$|\mu_{k_j}(E_j)| \geq |\mu_{k_j}(F_{k_j})| - |\mu_{k_j}(F_{k_{j+1}})| > \delta/2$. But this means

that $\{\mu_i\}$ is not uniformly strongly additive.

    We now obtain immediately the Nikodym Convergence Theorem for vector-valued measures ([38] III.7.4).

Theorem 9. Let Σ be a σ-algebra. Let $\mu_i : \Sigma \to Y$ be countably additive. If $\lim_i \mu_i(E) = \mu(E)$ exists for each $E \in \Sigma$, then

    (i) μ is countably additive and

    (ii) $\{\mu_i\}$ is uniformly countably additive.

Proof: (ii) follows from Theorem 6 (ii) and Lemma 8.

    For (i), let $\{E_j\}$ be a disjoint sequence from Σ. Then from the uniform countable additivity in (ii), we have

$$\mu(\bigcup_{j=1}^{\infty} E_j) = \lim_i \mu_i(\bigcup_{j=1}^{\infty} E_j) = \lim_i \lim_n \sum_{j=1}^{n} \mu_i(E_j) = \lim_n \mu(\bigcup_{j=1}^{n} E_j)$$

so that (i) holds.

    We now show that a version of the classical Vitali-Hahn-Saks Theorem (VHS) can be obtained from the results above. If $\lambda : \Sigma \to [0,\infty]$ is a non-negative (possibly infinite) set function and $\mu : \Sigma \to Y$, we say that μ is absolutely continuous with respect to λ (written $\mu << \lambda$) if for each $\epsilon > 0$ there exists $\delta > 0$ such that $|\mu(E)| < \epsilon$ whenever $E \in \Sigma$ and $\lambda(E) < \delta$. (That is, we use the ϵ-δ notion of absolute continuity.) If $\{\mu_i\}$ is a sequence of Y-valued set functions, $\{\mu_i\}$ is said to uniformly absolutely continuous with respect to λ if the δ in the definition above works for each μ_i. Using Theorem 9 we obtain a version of the VHS Theorem for countably additive measures from Theorem 9 ([38] III.7.2).

Theorem 10. Let Σ be a σ-algebra. Let $\mu_i : \Sigma \to Y$ be countably additive and let $\lambda : \Sigma \to [0,\infty]$ be countably additive. If $\mu_i << \lambda$ for each i and if $\lim_i \mu_i(E) = \mu(E)$ exists for each $E \in \Sigma$, then

    (i) $\mu << \lambda$

    (ii) $\{\mu_i\}$ is uniformly absolutely continuous with

respect to λ.

Proof: First note that if $\{E_j\}$ is a decreasing sequence from Σ with $E = \cap E_j$ and $\lambda(E) = 0$, then $\lim_j \mu_i(E_j) = 0$ uniformly in i. For $\{E_j \setminus E\}$ is a decreasing sequence with empty intersection so Theorem 9 and Lemma 8 imply that

$$\lim_j \mu_i(E_j \setminus E) = \lim_j (\mu_i(E_j) - \mu_i(E)) = \lim_j \mu_i(E_j) = 0$$

uniformly in i.

Now suppose that (ii) fails to hold. Then there is an $\epsilon > 0$ such that for each $\delta > 0$ there exist a positive integer k and $F \in \Sigma$ such that $|\mu_k(F)| \geq \epsilon$ and $\lambda(F) < \delta$. Put $\delta_1 = 1$, $k_1 = 1$ and $F_1 = \emptyset$. Then we may inductively define a subsequence $\{k_j\}$ and sequences $\{F_j\} \subseteq \Sigma$ and $\{\delta_j\} \subseteq R_+$ such that

$$\delta_{j+1} < \delta_j/2, \quad |\mu_{k_{j+1}}(F_{j+1})| \geq \epsilon, \quad \lambda(F_{j+1}) < \delta_j/2$$

and $|\mu_{k_{j+1}}(E)| < \epsilon/2$ whenever $\lambda(E) < \delta_{j+1}$. Set $E_j = \bigcup_{k=j}^{\infty} F_k$. Then, by the countable subadditivity,

(1)

$$\lambda(E_j \setminus F_j) \leq \lambda(E_{j+1}) \leq \sum_{k=j+1}^{\infty} \lambda(F_k) < \sum_{k=j}^{\infty} \delta_k/2 < \sum_{k=j}^{\infty} \delta_j/2^{k-j+1} = \delta_j$$

so that $\mu_{k_{j+1}}(E_{j+1} \setminus F_{j+1}) < \epsilon/2$. But $|\mu_{k_{j+1}}(F_{j+1})| =$

$|\mu_{k_{j+1}}(E_{j+1}) - \mu_{k_{j+1}}(E_{j+1} \setminus F_{j+1})| > \epsilon$ so that

(2) $|\mu_{k_j}(E_j)| \geq \epsilon/2$.

But $\{E_j\}$ is decreasing and if $E = \cap E_j$, then $\lambda(E) = 0$ by (1). By the observation in the first paragraph $\lim_j \mu_i(E_j) = 0$ uniformly in i. But this contradicts (2) and (ii) follows.

Condition (i) follows immediately from (ii).

The Nikodym Convergence Theorem, the Vitali-Hahn-Saks Theorem and the Nikodym Boundedness Theorem have also been generalized to set functions with domains which are not necessarily σ-algebras. In Schachermayer's treatise ([67]) these results (and others) are treated in detail, and we refer the reader to [67] for more complete details. To give a further application of our matrix methods we show that the Brooks-Jewett Theorem 6 is valid for bounded, finitely additive scalar set functions which are defined on quasi-σ-algebras. (In [67], Schachermayer refers to the conclusion of Theorem 6 as the Vitali-Hahn-Saks property instead of as the Brooks-Jewett property.)

Definition 11. A family A of subsets of a set S is called a quasi-σ-algebra if A is an algebra of sets such that each disjoint sequence $\{A_j\}$ from A has a subsequence $\{A_{k_j}\}$ such that $\cup A_{k_j} \in A$ ([28]).

This property (for Boolean algebras) is also considered in [40] under the name, subsequential completeness property; a slightly weaker form of this property is referred to in [67] as property (E) ([67] 4.2).

Theorem 12. Let A be a quasi-σ-algebra and let $\mu_i : A \to R$ be a bounded, finitely additive set function for each $i \in N$. If $\lim \mu_i(A) = \mu(A)$ exists for each $A \in A$, then $\{\mu_i\}$ is uniformly strongly additive.

Proof: First assume that each μ_i is countably additive. Consider the matrix $Z = [\mu_i(A_j)]$, where $\{A_j\}$ is a disjoint sequence from A. From the assumption that A is a quasi-σ-algebra and the countable additivity of the $\{\mu_i\}$, it follows that Z is a K-matrix. Hence, from the Basic Matrix Theorem, $\lim_{j} \mu_i(E_j) = 0$ uniformly for

$i \in \mathbb{N}$, and the result is established for this case.

Consider the case when the μ_i are only finitely additive. If the conclusion fails, we may assume (by passing to a subsequence if necessary) that there exist $\epsilon > 0$ and a disjoint sequence $\{A_j\}$ from A such that

(3)
$$| \mu_i(A_i) | > \epsilon.$$

Let Σ be the σ-algebra generated by A. Extend each μ_i to a bounded, finitely additive set function μ_i on Σ. By Drewnowski's Lemma there is a subsequence $\{A_{k_j}\}$ of $\{A_j\}$ such that each μ_i is countably additive on the σ-algebra, Σ_o, generated by the $\{A_{k_j}\}$. By the quasi-σ-algebra assumption, we may also assume that

$$A = \bigcup_{j=1}^{\infty} A_{k_j} \in A.$$

Put $A_A = \{B \cap A : B \in A\}$. Then $A_o = A_A \cap \Sigma_o$ is a quasi-σ-algebra of subsets of A. Each μ_i is countably additive on A_o and $\lim \mu_i(B) = \mu(B)$ exists for each $B \in A_o$. Therefore, by the first part, $\lim_j \mu_i(A_{k_j}) = 0$ uniformly in i. This contradicts (3).

In the terminology of [67], any quasi-σ-algebra has the Vitali-Hahn-Saks property (compare with [67] 4.3). Theorem 12 also gives an improvement of Proposition 1.B of [40] for quasi-σ-algebras of sets.

In an entirely similar fashion, the Nikodym Boundedness Theorem can also be shown to hold for quasi-σ-algebras (see [28] Theorem 1.7). In [67] it is shown that the Vitali-Hahn-Saks property always implies the Nikodym Boundedness Property, so we do not give the proof ([67] 2.5).

To further illustrate the applicability of the General Banach-Steinhaus Theorem we consider a result in classical summability due to Schur ([50] 7.1.6) Let $A = [a_{ij}]$ be an infinite matrix of real numbers. The matrix A is said to be of class (ℓ^{∞}, c) if

$\{\sum_{j=1}^{\infty} a_{ij} x_j\}_i$ is a convergent sequence for each sequence

$x = \{x_j\} \in \ell^{\infty}$. That is, if c is the space of all real sequences which converge, the formal matrix product Ax produces a sequence in c for each sequence $x \in \ell^{\infty}$. The classic Schur Theorem of summability gives necessary and sufficient conditions for a matrix A to be of class (ℓ^{∞}, c). In his treatment of the Schur Theorem, Maddox makes the remark that his proof is "classical" and that no functional analytic proof of the theorem seems to be known ([50], p. 168). We will give a proof below (of a more general summability result due to Hahn [39], [68]) based on the General Banach-Steinhaus Theorem; in the terminology of Maddox this might be considered a "functional analytic proof".

Recall that m_o is the subspace of ℓ^{∞} consisting of those sequences with finite range. We say that a matrix A is of <u>class</u> <u>(m_o,c)</u> if $Ax \in c$ for each $x \in m_o$. We give a characterization of matrices A of class (m_o, c).

We first require a lemma. Note that a sequence of real series $\sum_j |a_{ij}|$ converges uniformly in i if and only if for each $\epsilon > 0$ there exists N such that $|\sum_{j \in \sigma} a_{ij}| < \epsilon$ for each finite set σ with $\min \sigma \geqslant N$ and each i. (This is immediate since if the condition above is satisfied, then $\sum_{j=N}^{\infty} |a_{ij}| \leqslant 2\epsilon$ for each i ([61] I.1.2.).).

<u>Lemma 13</u>. Let $\sum_{j=1}^{\infty} |a_{ij}| < \infty$ for each i. If the series $\sum_j |a_{ij}|$ do not converge uniformly in i, there exist $\epsilon > 0$, a disjoint sequence of finite sets $\{\sigma_i\}$ and a subsequence $\{k_i\}$ such that $|\sum_{j \in \sigma_i} a_{k_i j}| > \epsilon$ for each i.

Proof: By the observation concerning uniform convergence above, if $\sum_j |a_{ij}|$ doesn't converge uniformly in i, there exists $\epsilon > 0$ such that for each i there are finite $\sigma_i \subseteq \mathbb{N}$ and a positive integer k_i with $\min \sigma_i \geqslant i$ and $|\sum_{j \in \sigma_i} a_{k_i j}| \geqslant \epsilon$.

Applying the observation above to $i_1 = 1$ implies that there exist a positive integer k_1 and a finite set σ_1 such that $|\sum_{j \in \sigma_1} a_{k_1 j}| > \epsilon$. There exists j_1 such that

$$(4) \qquad \sum_{j=j_1}^{\infty} |a_{ij}| < \epsilon \quad \text{for} \quad 1 \leqslant i \leqslant k_1 .$$

Applying the observation in the paragraph above to the integer $\max\{\max \sigma_1 + 1,\ j_1\} = i_2$ implies that there exist a positive integer k_2 and a finite set σ_2 with $\min \sigma_2 > i_2$ such that $|\sum_{j \in \sigma_2} a_{k_2 j}| \geqslant \epsilon$. Note that σ_1 and σ_2 are disjoint and by

$$(4), \qquad\qquad\qquad k_2 > k_1 .$$

Induction then produces the sequences in the conclusion of the lemma.

We now establish the summability result of Hahn.

Theorem 14. $A \in (m_o, c)$ if and only if

(i) $\sum_j |a_{ij}|$ converge uniformly in i

(ii) $\lim_i a_{ij} = a_j$ exists for each j.

Proof: We prove first that conditions (i) and (ii) imply that $A \in (\ell^{\infty}, c)$ so that in particular, $A \in (m_o, c)$. Let $\{x_j\} \in \ell^{\infty}$, where we may assume $|x_j| \leqslant 1$ for each j. Let $\epsilon > 0$. By (i), there exists N such that $m \geqslant n \geqslant N$ implies $|\sum_{j=n}^{m} a_{ij} x_j| < \epsilon$.

Thus, by (ii), $|\sum_{j=n}^{m} a_j x_j| \leqslant \epsilon$ so that the series $\sum_{j=1}^{\infty} a_j x_j$ converges. We now have,

(5)
$$|\sum_{j=1}^{\infty} a_{ij} x_j - \sum_{j=1}^{\infty} a_j x_j| \leqslant \sum_{j=1}^{N-1} |a_{ij} - a_j| + |\sum_{j=N}^{\infty} (a_{ij} - a_j) x_j|$$

$$\leqslant \sum_{j=1}^{N-1} |a_{ij} - a_j| + 2\epsilon .$$

Since the first term on the right hand side of (5) can be made less that ϵ for i large, this shows that

$$\lim_i \sum_{j=1}^{\infty} a_{ij} x_j = \sum_{j=1}^{\infty} a_j x_j , \text{ i.e., } A \in (\ell^{\infty}, c) .$$

If $A \in (m_o, c)$, then (ii) holds by taking $x = e_j \in m_o$.

Suppose that $A \in (m_o, c)$ and that (i) fails. Let the notation be as in Lemma 13 and consider the sequence $\{C_{\sigma_j}\}$ in m_o. Each row $\{a_{ij}\}_{j=1}^{\infty}$ of A induces a continuous linear functional R_i on m_o by $R_i x = \sum_{j=1}^{\infty} a_{ij} x_j$, where $x = \{x_j\}$, and $\lim_i R_i x = Rx$ exists. The sequence $\{C_{\sigma_j}\}$ is $\ell^1 - \mathcal{K}$ convergent in the duality between ℓ^1 and m_o. Therefore, by the General Banach-Steinhaus Theorem 3,

$$\lim_i R_i(C_{\sigma_j}) = \lim_i \sum_{k \in \sigma_j} a_{ik} = R(C_{\sigma_j}) = \sum_{k \in \sigma_j} a_k \text{ uniformly in } j. \text{ This}$$

implies $\lim_j \sum_{k \in \sigma_j} a_{ik} = 0$ uniformly in i and contradicts the conclusion in Lemma 13.

Theorem 14 has as an immediate corollary the summability result of Schur, i.e., $A \in (\ell^{\infty}, c)$ if and only if (i) and (ii) hold. We record these equivalences for later reference ([68]).

Corollary 15. The following are equivalent:

(a) $A \in (\ell^{\infty}, c)$

(b) $A \in (m_o, c)$

(c) (i) and (ii) of Theorem 14 hold.

A consequence of the Hahn-Schur summability result is another result due to Schur which states that a sequence in ℓ^1 is norm convergent if and only if it converges in the weak topology of ℓ^1 ([50] p. 170). Theorem 14 can be used to give an improved version of this result: a sequence in ℓ^1 converges in norm if and only if it is a Cauchy sequence with respect to the weak topology $\sigma(\ell^1, m_o)$ induced by the subspace m_o of ℓ^∞ .

To see that Theorem 14 yields this result, let $x_i = \{a_{ij}\}_{j=1}^\infty$ be a sequence in ℓ^1. Note that the sequence $\{x_i\}$ is a Cauchy sequence in the weak topology $\sigma(\ell^1, m_o)$ iff the matrix $[a_{ij}] = A$ belongs to (m_o, c). Thus, if the sequence $\{x_i\}$ is $\sigma(\ell^1, m_o)$-Cauchy, the matrix A satisfies conditions (i) and (ii) of Theorem 14. Let $x = \{a_j\}$ be the sequence given by (ii). We claim that x belongs to ℓ^1 and $\|x_i - x\|_1 \to 0$. Let $\epsilon > 0$. By (i), there exists N such that $\sum\limits_{j=N}^\infty |a_{ij}| < \epsilon$ for all $i \in \mathbb{N}$. Then from (ii) it follows that for $P > N$, $\sum\limits_{j=N}^P |a_j| \leqslant \epsilon$ and, hence, $\sum\limits_{j=N}^\infty |a_j| \leqslant \epsilon$. In particular, this shows that $x = \{a_j\}$ belongs to ℓ^1. Now, from (ii), there exists M such that $|a_{ij} - a_j| < \epsilon / N$ for $i \geqslant M$ and $1 \leqslant j \leqslant N - 1$. Hence, if $i \geqslant M$, we have

$$\sum_{j=1}^\infty |a_{ij} - a_j| \leqslant \sum_{j=1}^{N-1} |a_{ij} - a_j| + \sum_{j=N}^\infty |a_{ij}| + \sum_{j=N}^\infty |a_j| < 3\epsilon,$$

and, therefore, $x_i \to x$ in ℓ^1-norm. Of course, any norm convergent sequence converges with respect to $\sigma(\ell^1, m_o)$.

In Chapter 8, Theorem 1, we establish a version of this result for group-valued sequences and show that the abstract group version yields the result above (Corollary 8.2).

In Chapter 8, we also consider and establish versions of the summability results of Schur and Hahn for matrices whose entries are elements of a Banach space (Theorem 8.5). Another vector version of the summability result of Schur is given in Corollary 9.5 for bounded multiplier series in an F-space.

We conclude this section by indicating that another classical result in summability due to Toeplitz can also be derived from the Generalized Banach-Steinhaus Theorem. A matrix $A = [a_{ij}]$ is said to be of class (c,c) if $Ax \in c$ for every $x \in c$, and A is said to be regular if A is convergence preserving, i.e.,

$$\lim_i x_i = \lim_i \sum_{j=1}^{\infty} a_{ij} x_j \quad \text{for} \quad x = \{x_i\} \in c .$$ The result of Toeplitz

gives necessary conditions for a matrix A to be regular ([50] 7.1.3); the necessary conditions are also sufficient (Silverman).

Theorem 16. The matrix $A = [a_{ij}]$ is regular iff the following conditions hold:

(i) $\sup_i \sum_{j=1}^{\infty} |a_{ij}| < \infty$

(ii) $\lim_i a_{ij} = 0$ for each $j \in \mathbb{N}$

(iii) $\lim_i \sum_{j=1}^{\infty} a_{ij} = 1$.

Proof: The sufficiency of (i) - (iii), due to Siverman, are classical, and we do not repeat the proof (see [50] 7.1.3).

The necessity of (ii) and (iii) are easily obtained by putting $x = e_j$ and $x = (1,1,...)$. We show how the necessity of (i) can be established by utilizing the Generalized Banach-Steinhaus Theorem. Suppose that (i) does not hold. Then there exists a sequence, $\{\sigma_i\}$, of finite subsets of $\mathbb{N}$ such that $\max \sigma_i < \min \sigma_{i+1}$ and

(6) $$\sup_i | \sum_{j \in \sigma_i} a_{ij} | = \infty$$

Now each row $\{a_{ij}\}_{j=1}^{\infty}$ of A induces a continuous linear functional

R_i on c by $R_i x = \sum\limits_{j=1}^{\infty} a_{ij} x_j$, where $x = \{x_j\} \in c$, and $\lim\limits_i R_i x = Rx$

exists for each $x \in c$. The sequence $\{C_{\sigma_j}\}$ is $\ell^1 - \mathcal{K}$ convergent

in the duality between ℓ^1 and c so by the General Banach-Steinhaus Theorem 3,

$$\lim_i R_i(C_{\sigma_j}) = \lim_i \sum_{k \in \sigma_j} a_{ik} = R(C_{\sigma_j}) = \sum_{k \in \sigma_j} a_k$$

uniformly in j. Since $\lim\limits_j \sum\limits_{k \in \sigma_j} a_{ik} = 0$, we have $\lim\limits_i \sum\limits_{k \in \sigma_i} a_k = 0$,

and this contradicts (6).

6. Bilinear Maps and Hypocontinuity

In this section we consider bilinear maps and use the Basic
Matrix Theorem to derive a classical result of Mazur and Orlicz on
the joint continuity of separately continuous bilinear maps. We also
present some results pertaining to the hypocontinuity of bilinear
maps. Our initial results do not use the vector space structure of
the spaces involved and so are valid for normed groups. Many of the
results in this section are contained in [75], but the results presented here are slightly more general.

Let E, F and G be normed groups and let $b : E \times F \to G$ be a
biadditive map (i.e., the functions $b(x, \cdot) : F \to G$, $b(x, \cdot)(y)$
$= b(x, y)$, and $b(\cdot, y) : E \to G$, $b(\cdot, y)(x) = b(x, y)$, are additive for
each x and y).

For bilinear maps between topological vector spaces, Bourbaki
has introduced the notion of hypocontinuity which lies between separate continuity and joint continuity for bilinear maps. We define an
analogous notion for biadditive maps between topological groups and
establish some results on the hypocontinuity of such maps.

If $\aleph$ is a family of subsets of F, b is said to be
$\underline{\aleph\text{-hypocontinuous}}$ if for each neighborhood V of 0 in G and each
$N \in \aleph$, there is a neighborhood U of 0 in E such that
$b(U, N) \subseteq V$. Since the groups involved are metrizable, this is
equivalent to the following condition: if $x_i \to 0$ in E and $N \in \aleph$,
then $\lim b(x_i, y) = 0$ uniformly for $y \in N$.

Let $\sigma(E, F)$ be the weakest topology on E such that each additive map $b(\cdot, y)$, $y \in F$, is continuous. (The topology $\sigma(E, F)$ also
depends upon G and b, but this is suppressed in the notation and
should cause no difficulties.) $\sigma(F, E)$ is defined similarly. The
family of all $\sigma(E, F)-\aleph$ convergent sequences in E is denoted by

$K(E,F)$ ($K(F,E)$ is defined similarly).

We now state our first result; it is essentially a restatement of Theorem 5.3 for this particular setting.

Theorem 1. Let b be separately continuous. If $x_i \to 0$ in $\sigma(E,F)$ and $\{y_j\}$ is $\sigma(F,E)$- K convergent, then $\lim\limits_i b(x_i,y_j) = 0$

uniformly in j.

Proof: Consider the matrix $[b(x_i,y_j)]$. From the facts that b is separately continuous, $x_i \to 0$ in $\sigma(E,F)$ and $\{y_j\}$ is $\sigma(F,E)$- K convergent, it follows that this matrix is a K matrix. The Basic Matrix Theorem 2.2 gives the result.

We now present several hypocontinuity-type corollaries of Theorem 1.

Corollary 2. Let b be separately continuous. Then b is $K(F,E)$-hypocontinuous.

Proof: This follows immediately from Theorem 1 since if $x_i \to 0$ in E, then $x_i \to 0$ in $\sigma(E,F)$.

Let $K(F)$ be the family of all K convergent sequences in F (with respect to the original topology of F). Since the metric topology of F is stronger than $\sigma(F,E)$, $K(F) \subseteq K(F,E)$, and we have

Corollary 3. Let b be separately continuous. Then b is $K(F)$-hypocontinuous.

Corollary 3, which is a result concerning only the original topologies on E and F, now yields the following very interesting generalization of a classical result on joint continuity.

Corollary 4. Let b be separately continuous and let F be complete. Then b is continuous (i.e., jointly continuous).

Proof: Let $x_i \to 0$ $(y_i \to 0)$ in E (F). Since F is complete, $\{y_i\}$ is $\mathcal{K}$ convergent. By Corollary 3, $\lim_i b(x_i, y_j) = 0$ uniformly in j. In particular, $\lim_i b(x_i, y_i) = 0$, i.e., b is continuous.

For the case of metric linear spaces, Corollary 4 seems to be originally due to Mazur and Orlicz ([51]). The result seems to have been rediscovered many times and is sometimes attributed to Bourbaki ([22] 15.14) although the Bourbaki version requires the completeness of both of the (metric linear) spaces E and F. Baire category methods are usually utilized in the proofs of Corollary 4 found in the literature (cf [22] 15.14) as contrasted with the matrix methods used above. In the absence of completeness there do not appear to be any hypocontinuous-type of results of the nature of Corollaries 2 and 3. Of course, it should also be noted that the results above are valid for normed groups whereas the classical results pertaining to bilinear maps in the literature consider the case of (complete) metric linear spaces.

As an interesting application of Corollary 4, we consider the definition of a metrizable linear space due to Banach ([20], [46] 15.13). Banach developes a theory of metrizable linear spaces under the following set of axioms which (formally) appear to be weaker than those for a metric linear space. Let X be a vector space with a translation invariant metric d defined on X which satisfies the following properties:

(a) $t_n x \to 0$ for each $x \in X$ if $t_n \to 0$

(b) $t x_n \to 0$ for each scalar t if $x_n \to 0$

(c) X is complete with respect to the translation invariant metric d.

It can be shown that the vector space X under the topology
induced by d is actually a topological vector space ([46] 15.13) if
axioms (a) - (c) are satisfied. Using Corollary 4, it follows
immediately that if X satisfies only axioms (a) and (b), then X
is a toplogical vector space, i.e., the completeness in (c) is not
necessary. The proof of the fact that X is a topological vector
space given in [46] uses the Baire Category Theorem and, therefore,
will not yield the result in the absence of axiom (c).

A somewhat similar discussion concerning quasi-normed spaces is
carried out in [80] I.2. If X is a vector space, a quasi-norm on
X is a non-negative function $\| \ \|$ on X   satisfying:

(1) $\| x \| \geqslant 0$ and $\| x \| = 0$ iff $x = 0$

(2) $\| x + y \| \leqslant \| x \| + \| y \|$

(3) $\| - x \| = \| x \|$

and (a) and (b). It is shown in [80] I.2.2, that any quasi-normed
space is a topological vector space. The method of proof in [80]
relies on results of measure theory and should be contrasted with the
elementary matrix methods employed above.

We now consider the case when the spaces E, F and G are topological vector spaces, i.e., in our case metric linear spaces. In
this case, the analogues of Corollaries 2 and 3 are valid for the
family of $\mathcal{K}$ bounded sets. We let $b : E \times F \rightarrow G$ be a bilinear
map and retain the previous topological notation. Further, let
$\mathcal{K}_{\xi}(E,F)$ be the family of all $\sigma(E,F)- \mathcal{K}$ bounded subsets of E
(similarly for $\mathcal{K}_{\xi}(F,E)$), and let $\mathcal{K}_{\xi}(E)$ be the family of all $\mathcal{K}$
bounded subsets of E (similarly for $\mathcal{K}_{\xi}(F)$). We then have the
following analogue of Corollary 2 for $\mathcal{K}$ bounded sets.

Corollary 5. Let b be separately continuous. Then b is
$\mathcal{K}_{\xi}(F,E)$-hypocontinuous.

Proof: Let $x_j \to 0$ in E. It suffices to show that $\lim_i b(x_i, y_j) = 0$

uniformly in j for each $\sigma(F,E)$- $\mathcal{K}$ bounded sequence $\{y_j\} \subseteq F$.

If this condition fails to hold, there exist a $\sigma(F,E)$- $\mathcal{K}$

bounded sequence $\{y_j\}$ and an $\epsilon > 0$ such that for each i there

exist $k_i > i$, j_i with $|b(x_{k_i}, y_{j_i})| > \epsilon$. In particular, if

$i_1 = 1$, there exist k_1 and j_1 with $|b(x_{k_1}, y_{j_1})| > \epsilon$. Now since

$\lim_i b(x_i, y_j) = 0$ for each j, there exists i_2 such that

$|b(x_i, y_j)| < \epsilon$ for $i \geqslant i_2$ and $1 \leqslant j \leqslant j_1$. For i_2 there exist

$k_2 > i_2$ and j_2 such that $|b(x_{k_2}, y_{j_2})| > \epsilon$. Thus $j_2 > j_1$.

This construction can be continued to produce subsequences $\{k_i\}$ and

$\{j_i\}$ such that $|b(x_{k_i}, y_{j_i})| > \epsilon$. Thus, it suffices to show that

$\lim b(x_i, y_i) = 0$ for each sequence $x_i \to 0$ in E and each $\sigma(F,E)$-$\mathcal{K}$
bounded sequence $\{y_i\}$.

Pick a sequence of positive scalars $\{t_i\}$ such that $t_i \to \infty$
and $t_i x_i \to 0$. Then $\{(1/t_i)y_i\}$ is $\sigma(F,E)$- $\mathcal{K}$ convergent so Corollary 2 implies that $\lim b(t_i x_i, (1/t_i)y_i) = \lim b(x_i, y_i) = 0$, and
the result is established.

Since $\mathcal{K}(F) \subseteq \mathcal{K}(F,E)$, Corollary 5 has the following corollary
for the original topologies of E and F.

Corollary 6. Let b be separately continuous. Then b is
$\mathcal{K}(F)$-hypocontinuous.

In particular, if F is complete, Corollary 6 implies that b
is $\mathcal{K}(F)$-hypocontinuous when $\xi(F)$ is the family of all bounded
subsets of F ([22], [47]).

Note that there are no completeness or barrelledness assumptions
in either Corollary 5 or 6. The typical result in the literature

asserting some type of hypocontinuity has some sort of completeness or barrelledness assumption on the spaces involved (for example, [47] 40.2).

We next consider some results for sequences of bilinear mappings on the topological vector spaces E and F. Our results give a generalization of a result of Bourbaki for equicontinuous families of bilinear maps ([22] 15.14.3).

Theorem 7. Let $\{b_i\}$ be a sequence of separately continuous bilinear maps from $E \times F$ into G such that $\{b_i(x,y) : i\}$ is bounded for each $(x,y) \in E \times F$. Then $\{b_i\}$ is uniformly bounded on each product $A \times B \subseteq E \times F$ when

(i) A (B) is the range of a $\mathcal{K}$ convergent sequence in E (F).

(ii) A (B) is a $\mathcal{K}$ bounded subset of E (F).

Proof: Consider (i). Let $\{x_i\}$, $\{y_i\}$ be $\mathcal{K}$ convergent sequences in E and F, respectively.

Fix $y \in F$ and consider $b_i(\cdot,y) : E \to G$. This sequence of continuous linear operators is pointwise bounded and, therefore, uniformly bounded on $\mathcal{K}$ convergent sequences in E (Corollary 4.4). Hence, $\{b_i(x_j,y) : i,j\}$ is bounded for each $y \in F$, i.e., the family of operators $\{b_i(x_j, \cdot) : i,j\}$ is pointwise bounded on F. By Corollary 4.4, this family is uniformly bounded on $\mathcal{K}$ convergent sequences in F. Thus, $\{b_i(x_j,y_k) : i, j, k\}$ is bounded. This establishes (i).

Condition (ii) is established in a similar fashion using Corollary 4.4 (ii).

Theorem 7 can be viewed as a version of the Generalized Uniform Boundedness Principle (Theorem 4.2) for sequences of separately continuous bilinear maps. (In this regard also see Corollary 15 below.)

For complete spaces we have the following

Corollary 8. Let $\{b_i\}$ be as in Theorem 7. If E and F are complete, then $\{b_i\}$ is uniformly bounded on products of bounded subsets of E and F.

Proof: The corollary follows immediately from Theorem 7 since in complete spaces a set is $\mathcal{K}$ bounded iff it is bounded.

Compare Corollary 8 with 40.4.9 of [47].

If E, F and G are normed spaces, the conclusion in Corollary 8 implies that the sequence $\{b_i\}$ is equicontinuous since it is uniformly bounded on the product of the unit balls in E and F, and, thus, in the case of normed spaces, Corollary 8 gives the Bourbaki result in [22] 15.14.3.

We present an example which shows that the completeness assumption in Corollary 8 cannot be completely dropped.

Example 9. Define $b_i : \ell^\infty \times c_{oo} \to \mathbb{R}$ by $b_i(x,y) = \sum\limits_{j=1}^{i} x_j y_j$ where $x = \{x_i\}$, $y = \{y_i\}$. Then each b_i is separately continuous, and the sequence $\{b_i\}$ is pointwise bounded on $\ell^\infty \times c_{oo}$. If $e \in \ell^\infty$ is the constant sequence with a 1 in each coordinate, then

$b_i(e, f_i) = i$, where $f_i = \sum\limits_{j=1}^{i} e_j$, so that the sequence $\{b_i\}$ is not bounded on the product $\{e\} \times \{f_i : i \in \mathbb{N}\}$.

We next present another corollary which gives the general form of the Bourbaki result given in [22] 15.14.3.

Corollary 10. Let $\{b_i\}$ be as in Theorem 7. If E is complete, then $\lim\limits_{j} b_i(x_j, y_j) = 0$ uniformly in i for each sequence $\{x_j\}$ in E which converges to 0 and each sequence $\{y_j\}$ in F which is $\mathcal{K}$ convergent.

Proof: If the conclusion fails, there exists a neighborhood U of 0 in G such that for each i there are positive integers m_i, $n_i \geqslant i$ with $b_{m_i}(x_{n_i}, y_{n_i}) \notin U$.

For $i_1 = 1$ there exist m_1, n_1 such that $b_{m_1}(x_{n_1}, y_{n_1}) \notin U$. By Corollary 3 there exists $i_2 > n_1$ such that $j \geqslant i_2$ implies $b_i(x_j, y_j) \in U$ for $1 \leqslant i \leqslant m_1$. Now there exist $n_2 > i_2$ and m_2 such that $b_{m_2}(x_{n_2}, y_{n_2}) \notin U$. Note that $m_2 > m_1$ and $n_2 > n_1$. Thus this construction can be continued to produce two subsequences $\{m_i\}$ and $\{n_i\}$ such that $b_{m_i}(x_{n_i}, y_{n_i}) \notin U$.

Pick a sequence of positive scalars $\{t_i\}$ such that $t_i x_{n_i} \to 0$ and $t_i \to \infty$. The sequence $\{t_i x_{n_i}\}$ is $\mathcal{K}$ convergent by the completeness of E so by Theorem 7, $\{b_{m_i}(t_i x_{n_i}, y_{n_i})\}$ is bounded. Thus, $(1/t_i) b_{m_i}(t_i x_{n_i}, y_{n_i}) = b_{m_i}(x_{n_i}, y_{n_i}) \to 0$ contradicting the construction above.

Corollary 10 now yields the Bourbaki result ([22] 15.14.3) in its full generality as an immediate corollary.

Corollary 11. Let $\{b_i\}$ be as in Theorem 7. If E and F are complete, $\{b_i\}$ is equicontinuous.

Proof: Let $x_i \to 0$ in E and $y_i \to 0$ in F. By the completeness, $\{y_i\}$ is $\mathcal{K}$ convergent. By Corollary 10, $\lim_j b_i(x_j, y_j) = 0$ uniformly in i, i.e., $\{b_i\}$ is equicontinuous.

The example presented in Example 9 again shows that the completeness assumption in Corollary 11 cannot be completely dropped. (The calculations given below in Example 13 show that the sequence $\{b_i\}$ is not equicontinuous.)

Corollary 10 also yields an analogue of the Banach-Steinhaus Theorem for bilinear maps (see 4.5 and 5.1).

Corollary 12. Let $\{b_i\}$ be separately continuous and such that $\lim\limits_{i} b_i(x,y) = b(x,y)$ exists for each $(x,y) \in E \times F$. If E and F are complete, then $\{b_i\}$ is equicontinuous and b is continuous.

Proof: The equicontinuity follows from Corollary 11. If $x_j \to 0$ in E and $y_j \to 0$ in F, then $\{y_j\}$ is $\mathcal{K}$ convergent by the completeness of F so Corollary 10 implies that
$$\lim\limits_{j} b(x_j,y_j) = \lim\limits_{j} \lim\limits_{i} b_i(x_j,y_j) = 0, \quad \text{and} \quad b \text{ is continuous.}$$

The sequence in Example 9 shows that the completeness assumptions cannot be completely dropped.

We now consider some results on separate equicontinuity for bilinear maps. If $\{b_i\}$ is a sequence of bilinear maps from $E \times F \to G$, then $\{b_i\}$ is *right* (*left*) *equicontinuous* if the family of linear maps $\{b_i(x,\cdot) : i\}$ ($\{b_i(\cdot,y) : i\}$) is an equicontinuous family of linear maps from F into G (E into G) for each $x \in E$ ($y \in F$). The sequence is *separately equicontinuous* if it is both right and left equicontinuous.

We give an example of a sequence of bilinear maps which is left equicontinuous but not separately equicontinuous.

Example 13. Let $\{b_i\}$ be as in Example 9. Fix $y = \{y_j\} \in c_{oo}$ and assume that $y_j = 0$ for $j \geqslant n$. Then for $i \geqslant n$ $b_i(x,y) = \sum\limits_{j=1}^{n} x_j y_j$ for $x = \{x_j\} \in \ell^\infty$ so that $\|b_i(\cdot,y)\| \leqslant \sum\limits_{j=1}^{n} |y_j|$ and, therefore, $\{b_i(\cdot,y)\}$ is equicontinuous. Hence, $\{b_i\}$ is left equicontinuous. If $e \in \ell^\infty$ is the constant sequence with a 1 in each coordinate, then $||b_i(e,\cdot)|| = i$ so that $\{b_i(e,\cdot)\}$ is not equicontinuous. That is, $\{b_i\}$ is not right equicontinuous and, therefore, not separately equicontinuous.

We next observe that a left equicontinuous sequence of bilinear maps is pointwise bounded.

Proposition 14. Let $b_i : E \times F \to G$ be left equicontinuous. Then $\{b_i\}$ is pointwise bounded.

Proof: Let $(x,y) \in E \times F$ and $\{t_i\} \in c_o$. Then $t_i x \to 0$ in E so that $t_i b_i(x,y) = b_i(t_i x, y) \to 0$. Hence, $\{b_i(x,y)\}$ is bounded.

From Proposition 14, it follows that the conclusions of Theorem 7 and Corollaries 8, 10 and 11 hold for sequences of separately continuous, left equicontinuous bilinear maps.

Note that it follows from Corollary 4.5.(ii) that if $\{b_i\}$ is a pointwise bounded sequence of separately continuous bilinear maps and if E is complete, then $\{b_i\}$ is left equicontinuous. Combining this observation with Proposition 14 gives the following corollary.

Corollary 15. Let $\{b_i\}$ be separately continuous. If E is complete, the following are equivalent:

(i) $\{b_i\}$ is left equicontinuous

(ii) $\{b_i\}$ is pointwise bounded.

The sequence in Example 13 shows that the completeness assumption on E in Corollary 15 cannot be dropped. (Reverse the spaces c_{oo} and ℓ^∞ in Example 13).

Corollary 15 can be viewed as a version of the Uniform Boundedness Principle for sequences of separately continuous bilinear maps. (See Corollary 4.5.(ii).)

We now establish a generalized version of 40.2.2 of [47] for F-spaces. Since a left equicontinuous sequence of separately continuous bilinear functions is pointwise bounded, the corollary below follows immediately from Corollary 11. However, we give a direct proof based only on the Basic Matrix Theorem.

Corollary 16. Let $b_i : E \times F \to G$ be separately continuous and left equicontinuous. If E and F are complete, then $\{b_i\}$ is equicontinuous.

Proof: Let $\{x_j\}$ and $\{y_j\}$ converge to 0 in E and F, respectively.

Consider the matrix $[z_{ij}] = [b_i(x_i, y_j)]$. By the left equicontinuity $\lim_i z_{ij} = 0$ for each j. If $\{m_j\}$ is any increasing sequence of positive integers, by the completeness of F $\{m_j\}$ has a subsequence $\{n_j\}$ such that the series $\sum y_{n_j}$ is convergent. By the separate continuity, $\sum_j z_{in_j} = b_i(x_i, \sum_j y_{n_j})$, and the left equicontinuity then implies that $\lim_i \sum_j z_{in_j} = 0$. That is, the matrix $[z_{ij}]$ is a $\mathcal{K}$ matrix. By the Basic Matrix Theorem, $\lim_i z_{ii} = \lim b_i(x_i, y_i) = 0$, and the result follows.

For the case of complete metric linear spaces, Corollary 16 gives a generalization of 40.2.2 of [47] where it is shown that any separately equicontinuous sequence is equicontinuous. Note that the result above uses only the equicontinuity in one of the variables. The sequence in Example 13 shows that the completeness assumption cannot be completely eliminated.

Finally, we conclude this section by considering a result on uniform hypocontinuity of bilinear maps. Let $\{b_i\}$ be a sequence of bilinear maps from $E \times F$ into G and let $\mathcal{K}$ be a family of subsets of F. The sequence $\{b_i\}$ is said to be <u>$\mathcal{K}$-equihypocontinuous</u> if given any neighborhood U of 0 in G and any $N \in \mathcal{K}$, there is a neighborhood V of 0 in E such that $b_i(V,N) \subseteq U$ for all i. We now establish the analogues of Corollaries 3 and 5 for sequences of bilinear maps.

<u>Theorem 17.</u> Let $b_i : E \times F \to G$ be separately continuous and left equicontinuous. Then $\{b_i\}$ is

(i) $\mathcal{K}(F)$-equihypocontinuous and

(ii) $\mathcal{K}_{\sigma}(F)$-equihypocontinuous

<u>Proof:</u> For (i), it suffices to show that $\lim_i b_i(x_i, y_j) = 0$ uniformly in j when $x_i \to 0$ in E and $\{y_j\}$ is $\mathcal{K}$ convergent in F. Consider the matrix $[z_{ij}] = [b_i(x_i, y_j)]$. If $\{m_j\}$ is any subsequence, then by the $\mathcal{K}$ convergence of $\{y_j\}$ there is a subsequence $\{n_j\}$ of $\{m_j\}$ such that Σy_{n_j} converges to an element y of F.

By the separate continuity $\sum_j z_{in_j}$ converges and

$\lim_i \sum_j z_{in_j} = \lim_i b_i(x_i, y)$ exists by the left equicontinuity. Since

$\lim_i z_{ij} = 0$ for each j by the left equicontinuity, $[z_{ij}]$ is a $\mathcal{K}$

matrix. By the Basic Matrix Theorem, $\lim_i b_i(x_i, y_j) = 0$ uniformly

in j.

For (ii), it suffices to show that $\lim_i b_i(x_i,y_j) = 0$ uniformly in j when $x_i \to 0$ in E and $\{y_j\}$ is $\mathcal{K}$ bounded. If this condition fails to hold, then, as in the proof of Corollary 10, there exist a neighborhood U of 0 in G and subsequences $\{m_i\}, \{n_i\}$ such that $b_{m_i}(x_{m_i}, y_{n_i}) \notin U$. Pick a sequence of positive scalars $\{t_i\}$ such that $t_i \to \infty$ and $t_i x_{m_i} \to 0$. Then $\{(1/t_i)y_{n_i}\}$ is $\mathcal{K}$

convergent so by the first part,

$$\lim_i b_{m_i}(t_i x_{m_i}, (1/t_i)y_{n_i}) = \lim_i b_{m_i}(x_{m_i}, y_{n_i}) = 0.$$

This gives the desired contradiction.

For the case when F is a complete metric linear space, Theorem 17 yields 40.2.3(b) of [47] (or Proposition III.4.14 of [22]) as an immediate corollary. That is, if F is complete, then $\{b_i\}$ is equihypocontinuous with respect to the family of bounded subsets of F. Note that only the equicontinuity in one of the variables is used in this result. It is also worthwhile noting that Theorem 17 is valid with no completeness or barrelledness assumptions, and also that the non-locally convex case is covered in Theorem 17.

Finally, note that if both E and F are complete and $\{b_i\}$ is a sequence of pointwise bounded, separately continuous bilinear maps, then Corollary 15 and Theorem 17 imply that $\{b_i\}$ is equihypocontinuous with respect to the family of bounded subsets of F. (Compare with Exercise 40.8.c of [22].)

7. Orlicz-Pettis Theorems

In this section we consider the application of matrix methods to Orlicz-Pettis type theorems. The classical Orlicz-Pettis Theorem for normed spaces has been considered in section 3 as an application of a result on K convergent sequences (3.7). The classical Orlicz-Pettis Theorem guarantees that any series in a normed space which is subseries convergent for the weak topology is actually subseries convergent for the stronger norm topology. This result was originally proven by Orlicz for sequentially weakly complete spaces ([55]), and the first proof of the result for general normed spaces which was available to non-Polish speaking mathematicians was given by Pettis ([59]). An Orlicz-Pettis type of result is a theorem that asserts that a series which is subseries convergent in some topology is actually subseries convergent in some stronger topology. The literature abounds with such Orlicz-Pettis type results. We will present several such results which can be obtained by matrix methods, but no attempt will be made to give a complete survey of such Orlicz-Pettis type results. For historical remarks and extensive references to Orlicz-Pettis results, the reader is referred to [34] I.6 or [31].

Recall that a series Σx_j in the topological group E is <u>subseries</u> <u>convergent</u> if for each subsequence $\{x_{k_j}\}$, the subseries Σx_{k_j} converges in E. When there is more than one group topology on E, we have the following criteria for subseries convergence in both topologies.

<u>Theorem 1</u>. Let σ be a group topology on E which is weaker than the original topology of E and such that E has a basis at 0 which consists of sets which are closed for σ. If each series Σx_i

in E which is subseries convergent for σ satisfies $\lim|x_i| = 0$, then each series in E which is subseries convergent for σ is also subseries convergent for the original topology.

For the proof of Theorem 1, see [77], II.1 or [30], IV.1.1.

We first consider the classical Orlicz-Pettis Theorem. This result was derived in 3.8 as a corollary of a result on $\mathcal{K}$ convergence; here we give a simple direct proof based on the Basic Matrix Theorem.

Theorem 2. (Orlicz-Pettis) Let X be a normed space and let Σx_i be subseries convergent with respect to the weak topology. Then Σx_i is subseries convergent with respect to the norm topology.

Proof: By replacing X by the closed subspace spanned by the $\{x_i\}$, we may assume that X is separable.

Let $\{x_{n_i}\}$ be a subsequence of $\{x_i\}$ and set $s_i = \sum_{k=1}^{i} x_{n_k}$. We claim that $\{s_i\}$ is a norm Cauchy sequence. For this, let $\{p_i\}$ be an increasing sequence of positive integers and

$z_i = s_{p_{i+1}} - s_{p_i} = \sum_{k=p_i+1}^{p_{i+1}} x_{n_k}$, and note that it suffices to show

that $||z_i|| \to 0$. For each i, pick $z_i' \in X'$ such that $||z_i'|| = 1$ and $<z_i',z_i> = ||z_i||$. Since X is separable, by the Banach-Alaoglu Theorem, $\{z_i'\}$ has a subsequence $\{z_{k_i}'\}$ which converges weak* to an element $z' \in X'$.

Consider the matrix $[z_{ij}] = [<z_{k_i}',z_{k_j}>]$. By the weak* convergence of $\{z_{k_i}'\}$ and the weak subseries convergence of Σz_i, the matrix $[z_{ij}]$ is a $\mathcal{K}$ matrix. The Basic Matrix Theorem implies that $\lim z_{ii} = \lim ||z_{k_i}|| = 0$. Since the same argument can be applied to each subsequence of $\{z_i\}$, this shows that $||z_i|| \to 0$.

Since $\{s_i\}$ is norm Cauchy and weakly convergent, it follows that Σx_i is norm subseries convergent.

The locally convex version of the Orlicz-Pettis Theorem follows directly from Theorem 2 by simply applying Theorem 2 to each continuous semi-norm on the space. That is, if X is a locally convex space and if the series Σx_i is subseries convergent with respect to the weak topology of X, then the series is also subseries convergent with respect to the Mackey topology of X. In [78], Tweddle has established a very general form of the Orlicz-Pettis topology for locally convex spaces. We now give a result of somewhat the same nature as Tweddle's result which can be derived from the Basic Matrix Theorem.

Let E be a locally convex space and let $\mathcal{J}$ be the family of all series in E which are weak subseries convergent. Let G' be the vector space of all linear functionals y' on E such that $<y', \Sigma_i x_i> = \Sigma_i <y',x_i>$ for all $\Sigma x_i \in \mathcal{J}$. Clearly $E' \subseteq G'$. With respect to the natural duality between E and G', we have the following Orlicz-Pettis type result.

Theorem 3. If Σx_i is weak subseries convergent in the locally convex space E, then Σx_i is subseries convergent with respect to the topology of uniform convergence on $\sigma(G',E)$-Cauchy sequences in G'.

Proof: Let $\{y_j'\} \subseteq G'$ be $\sigma(G',E)$-Cauchy. Consider the matrix $[z_{ij}] = [<y_i',x_j>]$. Now $\lim_i z_{ij}$ exists since $\{y_j'\}$ is $\sigma(G',E)$-Cauchy. For any subsequence $\{k_j\}$, the sequence $\{\Sigma_j z_{ik_j}\} = \{<y_i', \Sigma_j x_{k_j}>\}$ converges since $\{y_i'\}$ is $\sigma(G',E)$-Cauchy. Thus, $[z_{ij}]$ is a $\mathcal{K}$ matrix. The Basic Matrix Theorem implies that

$\lim\limits_{j} z_{ij} = 0$ uniformly in i, i.e., $\lim\limits_{j} x_j = 0$ in the topology of uniform convergence on $\sigma(G', E)$-Cauchy sequences. Theorem 1 now gives the result.

Tweddle's Orlicz-Pettis result is somewhat different than Theorem 3 and in some sense is the strongest type of Orlicz-Pettis result that can be obtained for locally convex spaces. Tweddle shows that if Σx_i is weak subseries convergent, then Σx_i is subseries convergent with respect to the Mackey topology $\tau(E, G')$. Moreover, he shows that $\tau(E, G')$ is the strongest locally convex topology on E such that every element of $\mathcal{J}$ is subseries convergent.

Although Tweddle's result is more comprehensive than Theorem 3, Theorem 3 is general enough to have some interesting consequences. At this point we present an application of Theorem 3 to the scalar version of the Nikodym Convergence Theorem for countably additive measures. That is, we show that the Nikodym Convergence Theorem can be viewed as an Orlicz-Pettis result. Let Σ be a σ-algebra of subsets of a set S and let $\mu_i : \Sigma \to \mathbb{R}$ be countably additive for each $i \in \mathbb{N}$. Assume that $\lim\limits_{i} \mu_i(E) = \mu(E)$ exists for each $E \in \Sigma$. Let $S(\Sigma)$ be the vector space of all Σ-simple functions and let $ca(\Sigma)$ be all countably additive measures on Σ. Equip $S(\Sigma)$ with the weak topology $\sigma(S(\Sigma), ca(\Sigma)) = \sigma$ from the natural duality between $S(\Sigma)$ and $ca(\Sigma)$. If $\{E_j\}$ is a disjoint sequence from Σ, the series $\sum\limits_{j} C_{E_j}$ is σ-subseries convergent in $S(\Sigma)$. By Theorem 3, the series is subseries convergent in the toplogy of uniform convergence on $\sigma(ca(\Sigma), S(\Sigma))$ - Cauchy sequences. But, the sequence $\{\mu_i\}$ is $\sigma(ca(\Sigma), S(\Sigma))$ Cauchy by hypothesis. This means that the series $\sum\limits_{j} \mu_i(E_j)$ converge uniformly in i, i.e., the $\{\mu_i\}$ are uniformly countably additive. This is part of the conclusion in the Nikodym Convergence Theorem (Theorem 5.9 (ii)). The other part of the conclusion (Theorem 5.9 (i)) follows from this.

Notice that the analogue of Theorem 2 for the weak[*] topology is in general false. For example, the series Σe_j in ℓ^∞ is weak[*] subseries convergent but certainly not norm subseries convergent. Diestel and Faires have given necessary and sufficient conditions for an Orlicz-Pettis result to hold relative to the weak[*] and norm topologies ([33]). We derive the Diestel-Faires result in its full generality in section 10. At the present time we derive a somewhat less general result that follows quickly from the matrix methods employed up to this point.

Theorem 4. Let X be a normed space such that X contains no subspace isomorphic (topologically) to ℓ^1. If $\Sigma x_i'$ is weak[*] subseries convergent in X', then $\Sigma x_i'$ is norm subseries convergent.

Proof: For each i pick $x_i \in X$ such that $||x_i|| = 1$ and $<x_i', x_i> + (1/i) > ||x_i'||$. By a result of Rosenthal ([49] 2.e.5), $\{x_i\}$ has a subsequence $\{x_{k_i}\}$ which is weak Cauchy. Consider the matrix $[z_{ij}] = [<x_{k_j}', x_{k_i}>]$. Since $\{x_i\}$ is weak Cauchy and the series $\Sigma x_j'$ is weak[*] subseries convergent, $[z_{ij}]$ is a $\mathcal{K}$ matrix. The Basic Matrix Theorem implies that $\lim_i z_{ii} = 0$ so that $\lim ||x_i'|| = 0$. Theorem 1 now gives the result.

The Diestel-Faires result mentioned above replaces the condition that X contains no subspace isomorphic to ℓ^1 by the more general condition that X' contains no subspace isomorphic to ℓ^∞. We consider this more general result in section 10.

We next consider an Orlicz-Pettis result for compact operators which is due to Kalton ([44]). If X and Y are normed spaces, let $K(X,Y)$ be the space of compact operators from X into Y. Recall that the weak operator topology on $K(X,Y)$ is generated by the seminorms $T \to |<y', Tx>|$, $x \in X$, $y' \in Y'$ ([38] VI. 1.3). Kalton's result is given in

Theorem 5. Let $\Sigma\, T_i$ be subseries convergent in $K(X,Y)$ with respect to the weak operator topology. If X' contains no subspace isomorphic to ℓ^∞, then $\Sigma\, T_i$ is subseries convergent with respect to the norm topology of $K(X,Y)$.

Proof: First note that since each T_i is compact, it has separable range. Therefore, by replacing Y by the closed subspace spanned by the ranges of the $\{T_i\}$, we may assume that Y is separable.

Next, observe that the subseries convergence of $\Sigma\, T_j$ in the weak operator topology implies that the series $\Sigma\, T_j' y' = \Sigma\, y' T_j$ is weak* subseries convergent in X' for each $y' \in Y'$. Since X' contains no copy of ℓ^∞, the series $\Sigma\, T_j' y'$ is actually norm subseries convergent ([33] 1.2 or 10.10).

For each j pick $y_j' \in Y'$ such that $||y_j'|| = 1$ and $||T_j' y_j'|| + (1/j) \geqslant ||T_j'|| = ||T_j||$. Since Y is separable, the Banach-Alaoglu Theorem implies that $\{y_j'\}$ has a subsequence $\{y_{k_j}'\}$ which converges weak* to an element $y' \in Y'$. To avoid notational difficulties later, we assume $k_i = i$.

Now consider the matrix $[z_{ij}] = [T_j' y_i'] = [y_i' T_j]$. For each j, $\lim_i z_{ij} = \lim_i T_j' y_i'$ exists in norm by the compactness of T_j ([38] VI. 5.6). If $\{m_j\}$ is any subsequence, the series $\Sigma_j\, z_{im_j} = \Sigma_j\, T_{m_j}' y_i'$ converges in norm for each i. Moreover,

$$\lim_i \Sigma_j\, z_{im_j} = \lim_i \Sigma_j\, y_i' T_{m_j} = \lim_i\, y_i' \Sigma_j\, T_{m_j} = \lim_i\, (\Sigma_j\, T_{m_j})'\, y_i'$$

exists in norm by the compactness of $\Sigma_j\, T_{m_j}$ ([38] VI. 5.6). Hence, $[z_{ij}]$ is a $\mathcal{K}$ matrix. By the Basic Matrix Theorem,

$$\lim_i ||z_{ii}|| = \lim_i ||T_i' y_i'|| = 0 \quad \text{so} \quad \lim_i ||T_i'|| = \lim_i ||T_i|| = 0,$$

and Theorem 1 gives the result.

There are several observations concerning Theorem 5 that should be made. First, Theorem 5 is basically only valid for compact operators, such a result will not generally hold for $L(X,Y)$ unless $L(X,Y) = K(X,Y)$. For example, let X be a B-space with an unconditional Schauder basis $\{x_i, f_i\}$ (where f_i is the coefficient functional with respect to x_i). Now suppose X has the property that in the space $L(X,Y)$ any series which is subseries convergent in the weak operator topology is subseries convergent for the norm topology. Let P_k be the projection $P_k x = <f_k, x>x_k$. Then for any $T \in L(X,Y)$, the series $\sum_k TP_k$ is subseries convergent for the strong operator topology and, therefore, for the weak operator topology. If this series is norm subseries convergent, the sequence of compact operators $\{\sum_{k=1}^{i} TP_k\}$ is norm convergent to T so that $T \in K(X,Y)$. That is, $L(X,Y) = K(X,Y)$. In particular, if $X = Y$ and X has the property above, $L(X,X) = K(X,X)$, and X must be finite dimensional.

Next note that the assumption that X' contains no subspace isomorphic to ℓ^{∞} is actually necessary in the following sense. If the conclusion of Theorem 5 holds for all normed spaces Y, it must hold for the scalar field $\mathbb{R}$. But the weak operator topology of $K(X,\mathbb{R}) = X'$ is just the weak* topology of X'. Therefore, if the conclusion of the theorem holds, weak* subseries convergent series must be norm subseries convergent. By the Diestel-Faires result ([33]), X' cannot contain a subspace isomorphic to ℓ^{∞}. A more complete study of this situation is given in [32].

We next consider results concerning the topology of pointwise convergence in several of the classical function spaces. Let S be a compact Hausdorff space and let $C_G(S)$ be the space of all continuous functions from S into G, where G is a normed group. The

topology of pointwise convergence on $C_G(S)$ is the topology induced
by the family of quasi-norms, $f \to |f(t)|$, where $t \in S$ and $| |$ is
the quasi-norm which generates the topology of G. The topology of
uniform convergence on $C_G(S)$ is generated by the quasi-norm
$|f| = \sup\{|f(t)| : t \in S\}$. Concerning the function space $C_G(S)$, we
have the following Orlicz-Pettis result.

Theorem 6. Let $\{f_i\} \subseteq C_G(S)$. If Σf_i is subseries convergent with
respect to the topology of pointwise convergence, then Σf_i is
subseries convergent with respect to the topology of uniform convergence.

Proof: From Theorem 1, it suffices to show that $\{f_i\}$ converges to
0 in norm. For each i pick $t_i \in S_i$ such that

$$|f_i(t_i)| = \sup\{|f_i(t_i)| : t \in S\} = |f_i|.$$

We make the following claim: there exist a subsequence $\{t_{m_j}\}$
of $\{t_i\}$ and a $t \in S$ such that $\lim_i f_j(t_{m_i}) = f_j(t)$ for each

$j \in N$.

For this, let G^N be the space of all G-valued sequences
equipped with the quasi-norm

$$|g| = \sum_{i=1}^{\infty} |g_i| / 2^i (1 + |g_i|),$$

where $g = (g_1, g_2, \ldots) \in G^N$. Define $F : S \to G^N$ by
$F(s) = (f_1(s), f_2(s), \ldots)$. Note that F is continuous since each f_i
is continuous. The set $F(S)$ is compact and, moreover, since G^N
is a metric group, this set is sequentially compact. Thus, there is
a subsequence $\{F(t_{m_i}\}$ of $\{F(t_i)\}$ and an element $F(t) \in F(S)$ such
that $F(t_{m_i}) \to F(t)$ or, equivalently, $\lim_i f_j(t_{m_i}) = f_j(t)$ for each

j. This establishes the claim above.

Now suppose that $\{|f_i|\}$ doesn't converge to 0. Then we may assume that there exists $\epsilon > 0$ such that $|f_i| \geqslant \epsilon$ for each i. By the claim above there exist a subsequence $\{t_{m_i}\}$ of $\{t_i\}$ and $t \in S$ such that $\lim_i f_j(t_{m_i}) = f_j(t)$ for each j. To avoid subscripts assume that $m_i = i$. Note that $f_j(t) \to 0$ by the pointwise subseries convergence. Consider the matrix $Z = [z_{ij}] = [f_j(t_i) - f_j(t)]$. Note that the rows and columns of Z both converge to 0. Choose $\epsilon_{ij} > 0$ such that $\sum\limits_{i,j} \epsilon_{ij} < \infty$. By Lemma 2.1, there is an increasing sequence of positive integers $\{p_i\}$ such that $|z_{p_i p_j}| < \epsilon_{ij}$ for $i \neq j$. Let $f \in C_G(S)$ be such that $\sum\limits_{j=1}^{\infty} f_{p_j}(s) = f(s)$ for $s \in S$.

In particular, $\sum\limits_{j=1}^{\infty} f_{p_j}(t_{p_i}) = f(t_{p_i})$ for each i. By the claim above, there exist a subsequence $\{q_i\}$ of $\{p_i\}$ and an $x \in S$ such that $\lim\limits_i f_{p_j}(t_{q_i}) = f_{p_j}(x)$ for each j and $f(t_{q_i}) \to f(x)$. Moreover, $\lim\limits_i f_{p_j}(t_{q_i}) = f_{p_j}(t)$. Hence, $f_{p_j}(x) = f_{p_j}(t)$ for each j, and, therefore,

$$|f(x) - f(t)| = \left| \sum_j f_{p_j}(x) - \sum_j f_{p_j}(t) \right| \leqslant \sum_j |f_{p_j}(x) - f_{p_j}(t)| = 0.$$

By the triangle inequality, we have

$$|f_{q_i}(t_{q_i}) - f_{q_i}(t)| \leqslant \sum_{j \neq i} |f_{q_j}(t_{q_j}) - f_{q_j}(t)| + |f(t_{q_i}) - f(t)|$$

$$\leqslant \sum_{j=1}^{\infty} \epsilon_{ij} + |f(t_{q_i}) - f(x)| + |f(x) - f(t)|$$

$$= \sum_{j=1}^{\infty} \epsilon_{ij} + |f(t_{q_i}) - f(x)|.$$

Thus, $|f_{q_i}(t_{q_i}) - f_{q_i}(t)| \to 0$ and since $f_{q_i}(t) \to 0$, we have

$|f_{q_i}(t_{q_i})| \to 0$ which contradicts the assumption that $|f_i| \geqslant \epsilon$.

Hence, $|f_i| \to 0$ and the theorem is established.

The theorem above was established by Thomas for the case when G is a normed space ([77] II.4). Note that the matrix methods above easily handle the group case. The group case of the result above was established in [70] by using the Antosik-Mikusinski Diagonal Theorem.

In the case when S is metrizable, Thomas also shows that if a series Σf_i in $C_G(s)$ is subseries convergent with respect to the topology of pointwise convergence on a dense subset D of S, then it is also subseries convergent with respect to the topology of uniform convergence. The matrix method employed in the proof of Theorem 6 also yields this result.

Let D be a dense subset of S. The topology of pointwise convergence on D is the topology generated by the family of quasi-norms, $f \to |f(t)|$, $t \in D$. Concerning this topology, we have the Orlicz-Pettis result of Thomas ([77]).

Theorem 7. Let S be metrizable and let D be a dense subset of S. If Σf_i is subseries convergent with respect to the topology of point-wise convergence on D, then Σf_i is subseries convergent with respect to the topology of uniform convergence.

Proof: For each i, pick $t_i \in D$ such that $|f_i(t_i)| + (1/i) > |f_i|$. By the metrizability, $\{t_i\}$ has a subsequence $\{t_{k_i}\}$ which converges to a point $t \in S$. It is easily checked that the matrix $[z_{ij}] = [f_{k_j}(t_{k_i})]$ is a $\mathcal{K}$ matrix, and the proof is completed as in Theorem 6.

It should also be noted that the general case of Theorem 6 can be derived from the metric case established above in Theorem 7 by using a method of Thomas in [77], p. 183 (see also [38], VI.7.6).

Since the metric case of the theorem is very easily established, this method of Thomas represents an interesting contrast to the proof of Theorem 6 given above.

It is also worth noting that the scalar case of Theorem 6 contains the classical Orlicz-Pettis Theorem (Theorem 2) as a special case. Let X be a normed space and let Σx_i be subseries convergent in the weak topology. Let S be the unit ball of X' equipped with the weak* topology. Then S is compact and each x_i is a continuous function on S. The series Σx_i is subseries convergent with respect to the topology of pointwise convergence in $C(S)$, and, therefore, Σx_i is subseries convergent with respect to the topology of uniform convergence in $C(S)$. This implies that Σx_i is subseries convergent with respect to the norm topology.

It is of interest to note that the analogue of Theorem 6 is false for the space of bounded measurable functions; that is, the continuity of the functions in Theorem 6 is important. (The continuity was certainly utilized in the proofs of Theorems 6 and 7.) For let Σ be a σ-algebra of subsets of a set S and let $B(S,\Sigma)$ be the B space of all bounded real-valued functions on S which are Σ-measurable equipped with the sup-norm. If $\{E_j\}$ is a disjoint sequence of sets from Σ, the series ΣC_{E_j} is pointwise subseries

convergent but not norm subseries convergent. In particular, this holds for the space ℓ^∞. For the space $B(S,\Sigma)$ we have the following result for the topology of pointwise convergence.

Theorem 8. If Σf_i is subseries convergent in $B(S,\Sigma)$ with respect to the topology of pointwise convergence, then Σf_i is subseries convergent with respect to the Mackey topology $\tau(B(S,\Sigma), ca(\Sigma))$.

Proof: If σ is an infinite subset of $\mathbb{N}$, we write $\sum\limits_{i \in \sigma} f_i$ for the (pointwise) sum of the subseries $\sum\limits_{j} f_{m_j}$, where σ is ordered as the

subsequence $\{m_j\}$. If σ is finite, the meaning of $\sum\limits_{i\in\sigma} f_i$ is clear.

Define a family of countably additive scalar measures $\{\mu_t : t \in S\}$

on the power set $\mathcal{P}$ of $\mathbb{N}$ by $\mu_t(\sigma) = (\sum\limits_{i\in\sigma} f_i)(t)$. Since each

$\sum\limits_{i\in\sigma} f_i \in B(S,\Sigma)$, this family of measures is point-wise bounded on $\mathcal{P}$.

By the Nikodym Boundedness Theorem the family is uniformly bounded.

That is, the family $\{\sum\limits_{i\in\sigma} f_i : \sigma \subseteq \mathbb{N}\}$ is uniformly bounded in

$B(S,\Sigma)$.

If $\{m_j\}$ is any increasing sequence of positive integers, the

partial sums $\{\sum\limits_{j=1}^{k} f_{m_j}\}$ are uniformly bounded, so if ν is a count-

ably additive measure on Σ, the Bounded Convergence Theorem implies

that $\langle\nu, \sum\limits_{j} f_{m_j}\rangle = \sum\limits_{j} \langle\nu, f_{m_j}\rangle$. That is, the series $\sum f_j$ is subseries

convergent with respect to the topology $\sigma(B(S,\Sigma), ca(\Sigma))$. By the

version of the Orlicz-Pettis Theorem for locally convex spaces, $\sum f_i$

is subseries convergent for the Mackey topology $\tau(B(S,\Sigma), ca(\Sigma))$

([77]).

The results in [71] give a more detailed study of this situation

and, in a certain sense, show that Theorem 8 is the best result pos-

sible for the space $B(S,\Sigma)$.

We next consider the analogue of Theorem 6 for the ℓ^P-spaces.

Let $0 < p < \infty$. The space $\ell^P(G)$ is the space of all sequences

$f : \mathbb{N} \rightarrow G$ such that $\sum_i |f(i)|^P < \infty$. For $1 \leqslant p < \infty$, $|f|_p =$

$(\sum |f(i)|^P)^{1/P}$ defines a quasi-norm on $\ell^P(G)$, and for $0 < p < 1$,

$|f|_p = \sum |f(i)|^P$ defines a quasi-norm on $\ell^P(G)$.

The topology of pointwise convergence on $\ell^P(G)$ is the topology

generated by the quasi-norms, $f \rightarrow |f(i)|$ for $i \in \mathbb{N}$. Concerning

this topology we have the analogue of Theorem 6 for $\ell^P(G)$.

<u>Theorem</u> 9. Let $\Sigma \, f_i$ be subseries convergent with respect to the topology of pointwise convergence. Then $\Sigma \, f_i$ is subseries convergent with respect to the topology generated by $||_p$.

<u>Proof</u>: Consider the matrix $z_{ij} = (f_j(1), \ldots , f_j(i), 0, 0, \ldots) \in \ell^P(G)$. First $\lim_i z_{ij} = f_j$ in $\ell^P(G)$ since $f_j \in \ell^P(G)$. Next let $\{m_j\}$ be an increasing sequence of positive integers and let $\sum_{j=1}^{\infty} f_{m_j}$ represent the subseries sum with respect to the topology of pointwise convergence. Now

$$\sum_{j=1}^{\infty} z_{ij} = (\sum_{j=1}^{\infty} f_{m_j}(i), \ldots , \sum_{j=1}^{\infty} f_{m_j}(i), 0, \ldots) \text{ by the pointwise}$$

convergence and $\lim_i \sum_{j=1}^{\infty} z_{ij} = (\sum_{j=1}^{\infty} f_{m_j}(k))_{k=1}^{\infty}$ in $\ell^P(G)$ since $\Sigma \, f_{m_j} \in \ell^P(G)$. Thus, $[z_{ij}]$ is a $\mathcal{K}$ matrix. By the Basic Matrix Theorem, $\lim_j \lim_i |z_{ij}|_p = \lim_j |f_j|_p = 0$. Theorem 1 gives the result.

For the case when G is a normed space, Theorem 9 is due to Thomas ([77] II.4 and II.8). The group case was established in [70] by employing the Antosik-Mikusinski Diagonal Theorem.

Theorem 9 motivates the following abstract Orlicz-Pettis type result which is applicable to several of the classical sequence spaces as well as certain spaces having a Schauder basis ([74]).

<u>Theorem</u> 10. Let X be a metric linear space with a vector topology σ that is weaker than the original metric topology. Suppose there exists a sequence of linear operators $T_i : X \rightarrow X$ such that
   (i) each T_i is $\sigma - ||$ continuous
   (ii) $\lim T_i x = x$ for each $x \in X$ (i.e., $\{T_i\}$ converges to the identity operator in the strong operator topology). If the series

Σx_i is σ-subseries convergent, then Σx_i is $||$-subseries convergent.

<u>Proof</u>: Consider the matrix $[z_{ij}] = [T_i x_j]$. From (ii), it follows that $\lim_i z_{ij} = x_j$ for each j. From (i), it follows that for any increasing sequence of positive integers $\{m_j\}$, $\sum_j z_{im_j}$ converges to $T_i(\sum_j x_{m_j})$ (where $\sum_j x_{m_j}$ is the σ-limit of the subseries). Hence, (ii) implies that $\lim_i \sum_j z_{im_j} = \sum_j x_{m_j}$ exists. Thus, $[z_{ij}]$ is a $\mathcal{K}$ matrix. The Basic Matrix Theorem implies that $\lim_j \lim_i T_i x_j = \lim_j x_j = 0$ (in the metric topology). Theorem 1 now yields the result.

It is easy to see that the topological vector space version of Theorem 9 is a corollary of Theorem 10 by taking σ to be the topology of pointwise convergence and by defining $T_i : \ell^p(X) \to \ell^p(X)$ by $T_i(x_1, x_2, \ldots) = (x_1, \ldots x_i, 0, \ldots)$. The topology of pointwise convergence on the sequence spaces $c_0(X)$ and $c(X)$ can be treated in a similar fashion.

A B-space with a Schauder basis can also be treated by using Theorem 10. Let X be a B-space with a Schauder basis $\{x_i, f_i\}$ (f_i is the coordinate functional relative to x_i). Let Γ be the subspace of X' spanned by the $\{f_i\}$ and let σ be the weak topology $\sigma(X, \Gamma)$. In the case of ℓ^p, c_0 or c, this topology is just the topology of pointwise convergence.

Define $T_i : X \to X$ by $T_i x = \sum_{k=1}^{i} <f_k, x> x_k$. Then each T_i is $\sigma - ||\ ||$ continuous and $\{T_i\}$ converges to the identity operator in the strong operator topology. By Theorem 10 any series Σy_i in X which is σ-subseries convergent is also norm-subseries convergent.

We conclude this section by establishing an abstract type of

Orlicz-Pettis result which was presented in [70]. Let E, F and G be normed groups and let $b : E \times F \to G$ be a biadditive map. As in section 6, we let $\sigma(E,F)$ be the weakest topology on E such that each of the additive maps $\{b(\cdot,y) : y \in F\}$ is continuous. (The topology $\sigma(F,E)$ is defined similarly.) Again the topology $\sigma(E,F)$ also depends on the space G and the map b, but the notation should cause no difficulties.

Let $\mathbb{C}_F$ be the family of all $\sigma(F,E)$-Cauchy sequences in F. We let $\mathbb{C}(E,F)$ be the topology on E of uniform convergence on elements of the family $\mathbb{C}_F$ of $\sigma(F,E)$-Cauchy sequences in F, i.e., a net $\{x_v\}$ in E converges to 0 in the topology $\mathbb{C}(E,F)$ iff $\lim b(x_v,f_i) = 0$ uniformly in i for each $\{f_i\} \in \mathbb{C}_F$. The topology $\mathbb{C}(E,F)$ is generated by the quasi-norms $|x|_\alpha = \sup|b(x,f_i)|$, where $\alpha = \{f_i\}$ ranges over the family of all $\sigma(F,E)$-Cauchy sequences.

Concerning these topologies, we have the following Orlicz-Pettis result.

Theorem 11. If $\Sigma \, x_i$ is subseries convergent with respect to $\sigma(E,F)$, then $\Sigma \, x_i$ is subseries convergent with respect to $\mathbb{C}(E,F)$.

Proof: Let $\{f_i\}$ be $\sigma(F,E)$-Cauchy. Consider the matrix $[z_{ij}] = [b(x_j,f_i)]$. Since $\{f_i\}$ is $\sigma(F,E)$-Cauchy, $\lim_i z_{ij}$ exists.

If $\{m_j\}$ is an increasing sequence of positive integers, then $\sum_j z_{im_j} = b(\Sigma x_{m_j}, f_i)$, where Σx_{m_j} is the $\sigma(E,F)$-sum of the subseries. Thus, $\lim_i \sum_j z_{im_j}$ exists since $\{f_i\}$ is $\sigma(F,E)$-Cauchy.

Hence, $[z_{ij}]$ is a $\mathcal{K}$ matrix. The Basic Matrix Theorem implies that $\lim_j b(x_j,f_i) = 0$ uniformly in i, i.e., $x_j \to 0$ in $\mathbb{C}(E,F)$.

Theorem 1 now gives the result.

Theorem 11 is established in [70] by employing the Antosik-

Mikusinski Diagonal Theorem, and it is shown in [70] that a large number of known Orlicz-Pettis results can be derived from Theorem 11 as corollaries. For example, Theorems 6 and 9 can be obtained directly from Theorem 11. We refer the reader to [70] for details.

8. The Schur and Phillips Lemma

In this section we discuss the classical lemmas of Schur and
Phillips and show how these results can be treated by the matrix
methods developed in section 2. One version of the classical Schur
lemma asserts that a sequence in ℓ^1 converges weakly iff it converges strongly. This result and some of its more general forms have
found many applications in functional analysis; for example, many of
the proofs of the Orlicz-Pettis Theorem, including the original proof
of Pettis ([59]), use the Schur lemma in some form. Similarly,
Phillips' lemma has many applications in both measure theory and
functional analysis; for example, the original application of
Phillips showed that there is no continuous projection of ℓ^∞ onto
c_0 ([60] 14.4.0). Both of these results have been generalized to
various abstract settings. For example, Brooks has given Banach
space versions for both the Schur and Phillips' lemmas ([23]) and
Robertson has given a group version of the Schur lemma ([62]). In
this section we present group-valued versions of both the Schur and
Phillips lemmas. As an application of our general Schur lemma, we
give a generalization of another classical result in summability
theory which is also due to Schur.

We first establish our generalization of the Schur lemma and
then indicate how this result can be legitimately viewed as a generalization of the classical Schur lemma.

Throughout this section, G will denote a normed group. If σ
is an infinite subset of $\mathbb{N}$ and if $\sum_j x_j$ is subseries convergent in

G, we write $\sum_{j \in \sigma} x_j$ for the sum of the series $\sum_{j=1}^\infty x_{n_j}$, where the
elements of σ are arranged in the subsequence $\{n_j\}$. If $\sigma \subset \mathbb{N}$ is
finite, the meaning of $\sum_{j \in \sigma} x_j$ is clear.

Theorem 1. Let $x_{ij} \in G$ for $i, j \in \mathbb{N}$. Assume that the rows of the matrix $[x_{ij}]$ are subseries convergent and $\lim_i x_{ij} = x_j$ exists for each j. If $\{\sum_{j \in \sigma} x_{ij}\}$ is convergent in G for each $\sigma \subseteq \mathbb{N}$, then

   (i)   the series Σx_j is subseries convergent and

   (ii) $\lim_i \sum_{j \in \sigma} x_{ij} = \sum_{j \in \sigma} x_j$ uniformly for $\sigma \subseteq \mathbb{N}$.

Proof: First we show that the sequence $\{\sum_{j \in \sigma} x_{ij}\}$ satisfies a Cauchy condition uniformly with respect to $\sigma \subseteq \mathbb{N}$. If this is not the case, there is a $\delta > 0$ and a subsequence $\{n_i\}$ such that

$$(1) \qquad \sup_\sigma \left| \sum_{j \in \sigma} (x_{n_{i+1}j} - x_{n_i j}) \right| > \delta.$$

Set $z_{ij} = x_{n_{i+1}j} - x_{n_i j}$ and $m_1 = 1$. By (1) there exists a finite σ_1 such that $\left| \sum_{j \in \sigma_1} z_{m_1 j} \right| > \delta$. Set $N_1 = \max \sigma_1$. Since the columns of $\{z_{ij}\}$ go to 0, there is an $m_2 > m_1$ such that $\sum_{j=1}^{N_1} |z_{ij}| < \delta/2$ for $i \geqslant m_2$. Again by (1), there is a finite σ_2 such that $\left| \sum_{j \in \sigma_2} z_{m_2 j} \right| > \delta$. Set $\tau_1 = \sigma_1$ and $\tau_2 = \sigma_2 \setminus \{j : 1 \leqslant j \leqslant N_1\}$. Note τ_1 and τ_2 are disjoint with

$$\max \tau_1 < \min \tau_2 \quad \text{and} \quad \left| \sum_{j \in \tau_2} z_{m_2 j} \right| \geqslant \left| \sum_{j \in \sigma_2} z_{m_2 j} \right| - \sum_{j=1}^{N_1} |z_{m_2 j}| > \delta/2.$$

Continuing this construction produces a subsequence $\{m_i\}$ and disjoint finite sets $\{\tau_i\}$ satisfying

$$(2) \qquad \left| \sum_{j \in \tau_i} z_{m_i j} \right| > \delta/2.$$

Now consider the matrix $[y_{ij}] = [\sum_{k \in \tau_j} z_{m_i k}]$. The columns of

$\{y_{ij}\}$ go to 0, and if $\{p_j\}$ is any subsequence, $\sum_{j=1}^{\infty} y_{ip_j}$ is a subseries of the series $\sum_j z_{m_i j}$ Thus, the matrix $\{y_{ij}\}$ is a $\mathcal{K}$ matrix. Hence, $\lim_i y_{ii} = 0$. But this contradicts (2).

We now establish (i) and also

$$(3) \qquad \sum_{j=1}^{\infty} x_{n_j} = \lim_i \sum_{j=1}^{\infty} x_{in_j},$$

for any subsequence $\{n_j\}$. Let $\epsilon > 0$. By what we have just established, there exists an N such that

$$(4) \qquad |\sum_{j\in\sigma} (x_{ij}-x_{kj})| < \epsilon/3 \quad \text{for} \quad i,k \geqslant N.$$

Hence, for each M and $k > N$, $|\sum_{j=1}^{M} (x_{n_j}-x_{kn_j})| \leqslant \epsilon/3$. Thus

$$(5) \qquad |\sum_{j=1}^{M} x_{n_j} - \sum_{j=1}^{\infty} x_{kn_j}| \leqslant |\sum_{j=1}^{M} (x_{n_j} -x_{kn_j})| + |\sum_{j=M+1}^{\infty} (x_{kn_j} -x_{Nn_j})|$$

$$+ |\sum_{j=M+1}^{\infty} x_{Nn_j}| < 2\epsilon/3 + |\sum_{j=M+1}^{\infty} x_{Nn_j}|,$$

and the last term on the right hand side of (5) is small for M large. This establishes (3).

From (3) and the uniform Cauchy condition (4), it follows that $\lim_i \sum_{j\in\sigma} x_{ij} = \sum_{j\in\sigma} x_j$ uniformly for $\sigma \subseteq N$ and the proof is complete.

Theorem 1 generalizes the version of the Schur lemma for B-spaces as given by Brooks in [23], Corollary 2. Brooks methods do not generalize to non-locally convex spaces or groups since they depend upon duality methods. Robertson has established a (more general) form of Theorem 1 in [62]; his methods are quite different than the matrix methods employed above and depend heavily on Baire category methods.

We now derive the classical Schur lemma from Theorem 1 thus indicating that Theorem 1 can be viewed as a legitimate generalization of the classical Schur lemma.

Corollary 2. (Schur) Let $\{t_{ij}\}$ be a real matrix. If $\lim\limits_{i} \sum\limits_{j \in \sigma} t_{ij}$

exists for each $\sigma \in \mathbb{N}$ and if $t_j = \lim\limits_{i} t_{ij}$, then $\{t_j\} \in \ell^1$

and $\lim\limits_{i} \sum\limits_{j=1}^{\infty} |t_{ij} - t_j| = 0$.

Proof: Let $\epsilon > 0$. By Theorem 1, for large i,

$|\sum\limits_{j \in \sigma} (t_{ij} - t_j)| < \epsilon$ for $\sigma \subseteq \mathbb{N}$. But then for such i,

$\sum\limits_{j=1}^{\infty} |t_{ij} - t_j| \leq 2\epsilon$ ([61] 1.1.2).

The usual statement of the Schur lemma has $t_j = 0$ in Corollary 2 ([79] 1.3.1). This slightly more general form has been given by Brooks and Mikusinski in [25]. The function space interpretation of Corollary 2 is the following. Let $x_i = \{t_{ij}\} \in \ell^1$ for each i and let m_0 be the subspace of ℓ^∞ which consists of the sequences with finite range. The hypothesis in Corollary 2 is just that the sequence $\{x_i\}$ is a Cauchy sequence in the topology $\sigma(\ell^1, m_0)$. The conclusion is then that the sequence $\{x_i\}$ is actually norm convergent in ℓ^1. In particular, this implies that any weakly convergent sequence in ℓ^1 is norm convergent; this is one way that the Schur lemma is sometimes stated ([79] 14.4.7).

For use in section 9 we also note that the conclusion in Theorem 1 can be strengthened somewhat when the matrix has values in a Banach space. Recall that a series Σx_i in a topological vector space is said to be bounded multiplier convergent if for each sequence $\{t_j\}$ of bounded scalars, the series $\Sigma t_i x_i$ is convergent. In a B-space

subseries convergence and bounded multiplier convergence are equivalent ([63] III.6.5). For series in B-spaces which satisfy the
conditions in Theorem 1, we have the following uniform bounded
multiplier result.

Corollary 3. Let X be a B-space and $x_{ij} \in X$ satisfy the
hypothesis of Theorem 1. Then

$$\lim_i \sum_{j=1}^{\infty} t_j x_{ij} = \sum_{j=1}^{\infty} t_j x_j \quad \text{uniformly for } \{t_j\} \in \ell^{\infty} \text{ with } ||\{t_j\}|| \leqslant 1.$$

Proof: Let $\epsilon > 0$. By Theorem 1, there exists N such that $i \geqslant N$
implies $||\sum_{j \in \sigma} (x_{ij} - x_j)|| < \epsilon$ for $\sigma \subseteq \mathbf{N}$. Thus, for $x' \in X'$ and

$||x'|| \leqslant 1$, $|\sum_{j \in \sigma} < x', x_{ij} - x_j > | < \epsilon$ for $i \geqslant N$ and $\sigma \subseteq \mathbf{N}$. This

implies that

$$(6) \qquad \sum_{j=1}^{\infty} | < x', x_{ij} - x_j > | \leqslant 2\epsilon \quad \text{for } i \geqslant N, \quad ||x'|| \leqslant 1$$

([61] 1.1.2). Now for $\{t_j\} \in \ell^{\infty}$ with $|t_j| \leqslant 1$, we have from (6)

$$||\sum_{j=1}^{\infty} t_j (x_{ij} - x_j)|| = \sup\{| < x', \sum_{j=1}^{\infty} t_j (x_{ij} - x_j) > | : ||x'|| \leqslant 1\}$$

$$\leqslant \sup\{\sum_{j=1}^{\infty} |t_j| \ | < x', x_{ij} - x_j > | : ||x'|| \leqslant 1\} \leqslant 2\epsilon$$

for $i \geqslant N$.

Corollary 3 has the following interesting operator interpretation. Each subseries convergent series $\sum_j x_{ij}$ induces a bounded

linear operator $T_i : \ell^{\infty} \to X$ by $T_i\{t_j\} = \sum_j t_j x_{ij}$. The hypothesis

of Theorem 1 implies that the sequence $\{T_i\}$ is a Cauchy sequence
for the topology of pointwise convergence on the subspace m_o of ℓ^{∞}.
The conclusion of Corollary 3 is then that there exists a bounded

linear operator T which is induced by the subseries convergent
series Σx_j and the sequence $\{T_i\}$ actually converges to T in the
uniform operator or norm topology of $L(\ell^\infty, X)$. With this interpretation, Corollary 3 can be viewed as a vector version of the
classical Schur lemma (Corollary 2).

We next use Theorem 1 to derive a generalization of a classical
summability result which is also due to Schur (see Corollary 5.13).
Recall that a scalar matrix $A = [a_{ij}]$ is said to be of class
(ℓ^∞, c) $((m_0, c))$ if for each $x = \{t_j\} \in \ell^\infty$ $(\{t_j\} \in m_0)$, the sequence $\{\sum_j a_{ij} t_j\}_i$ is convergent, i.e., if for each $x \in \ell^\infty$ $(x \in m_0)$,
the formal matrix product Ax is well-defined and produces a
sequence belonging to c (§ 5 or [50] § 7). The classical summability result of Hahn and Schur gives necessary and sufficient
conditions for a matrix A to be of class (ℓ^∞, c) or (m_0, c) (Cor.
5.15, [68] or [50] 7.6). We use Theorem 1 to give a generalization
of this sufficient condition to matrices with values in a group.
Namely, we have the following corollary to Theorem 1.

Corollary 4. Let x_{ij} be as in Theorem 1. Then the series $\Sigma_j x_{ij}$
are unordered uniformly convergent in the sense that if $\epsilon > 0$, there
exists N such that $|\sum_{j \in \sigma} x_{ij}| < \epsilon$ for all i when min $\sigma \geqslant$ N.

Proof: By the uniform Cauchy condition of Theorem 1, there is an M
such that $|\sum_{j \in \sigma} (x_{ij} - x_{kj})| < \epsilon/2$ for i, k > M and $\sigma \subseteq$ N. Since each
$\sum_j x_{ij}$ is subseries convergent, there is an N such that
$|\sum_{j \in \sigma} x_{ij}| < \epsilon/2$ whenever $1 \leqslant i \leqslant M$ and min $\sigma \geqslant$ N. Hence, for
min $\sigma \geqslant$ N and $i \geqslant$ M, $|\sum_{j \in \sigma} x_{ij}| \leqslant |\sum_{j \in \sigma} (x_{ij} - x_{Mj})| + |\sum_{j \in \sigma} x_{Mj}| < \epsilon$

and the result follows.

For scalar matrices $A = [a_{ij}]$, the conclusion of Corollary 4 is equivalent to the sufficient condition (i) in Theorem 5.14. For, if the conclusion of Corollary 4 holds for A, then $\sum_{j \geqslant N} |a_{ij}| \leqslant 2\epsilon$ ([61] I.1.2) and this is just condition 5.14 (i) in the classical Schur summability result. Thus, Theorem 1 and Corollary 4 yield another proof of the Schur summability result given in 5.15.

We can also use the results in Theorem 1 and Corollaries 3 and 4 to consider a vector version of the classical summability results above. Let X be a B-space and let $c(X)$ be the space of all X-valued sequences $\{x_j\}$ which are convergent. If $A = [x_{ij}]$ is an infinite matrix with values in X, then A is said to be in the class $(m_0, c(X))$ $((\ell^\infty, c(X)))$ if for each $y = \{t_j\} \in m_0$ $(\{t_j\} \in \ell^\infty)$, the sequence $\{\sum_j t_j x_{ij}\}_i$ belongs to $c(X)$, i.e., if the formal matrix product Ay is well-defined for each $y \in m_0$ $(y \in \ell^\infty)$ and produces a sequence in $c(X)$. Using the results above, we have the following vector summability result which gives a vector version of the classical Hahn-Schur summability results (5.15).

Theorem 5. Let $A = [x_{ij}]$ be an infinite matrix whose rows are subseries convergent. The following are equivalent:

(a) $A \in (\ell^\infty, c(X))$

(b) $A \in (m_0, c(X))$

(c) (i) the series $\sum_j x_{ij}$ are unordered uniformly convergent

(see Cor. 4) and

    (ii) $\lim_i x_{ij} = x_j$ exists for each j.

(d) $\lim_i \sum_{j \in \sigma} x_{ij} = \sum_{j \in \sigma} x_j$ uniformly for $\sigma \subseteq \mathbb{N}$.

Proof: That (a) implies (b) is clear and (b) implies (a) by Corollary 3. Corollary 4 shows that (b) implies (c)(i) and clearly (b) implies (c)(ii) by considering $e_j \in m_o$.

We next show (c) implies (d). Let $\epsilon > 0$. By (i) there exists N such that $\| \sum_{j \in \sigma} x_{ij} \| < \epsilon$ for all i and $\min \sigma \geqslant N$. Let $\sigma \subseteq N$ and let $\sigma(n) = \sigma \cap \{ j : j \geqslant n \}$. If $\sigma = \{ n_i : n_i < n_{i+1}, i \in N \}$ and if $q > p \geqslant N$, then $\| \sum_{i=p}^{q} x_{n_i} \| \leqslant \epsilon$ by (ii). Hence, since X is a B-space, the subseries Σx_{m_i} converges. That is, the series Σx_i is subseries convergent, and the sum $\sum_{i \in \sigma} x_i$ exists. Then

$$\| \sum_{j \in \sigma} (x_{ij} - x_j) \| \leqslant \sum_{j=1}^{N} \| x_{ij} - x_j \| + \| \sum_{j \in \sigma(N)} x_{ij} \| + \| \sum_{j \in \sigma(N)} x_j \| < 3\epsilon$$

for $i \geqslant M$. Hence (d) holds.

That (d) implies (b) is clear.

In a very similar fashion, one can formulate and consider analogues of the classical summability results of Hahn and Schur for matrices whose entries are bounded linear operators between metric linear spaces. Such analogues of the classical summability results are obtained in [76] and are based on the matrix methods of these notes.

We conclude this section with a discussion of the Phillips lemma. First, we give a statement of the classical scalar version of the Phillips lemma ([60], [79] 14.4.4).

Theorem 6. (Phillips [60]) Let $\mathcal{P}$ be the power set of N and let

$\mu_i : \mathcal{P} \to \mathbb{R}$ be bounded and finitely additive. If $\lim_i \mu_i(E) = \mu(E)$

exists for each $E \subseteq \mathbb{N}$, then $\lim_i \sum_{j=1}^{\infty} |\mu_i(j) - \mu(j)| = 0$.

Theorem 6 is often stated in this form with $\mu(E) = 0$ ([60], [79]).

Theorem 6 has the following function space interpretation. The dual of ℓ^{∞} is the space, ba, of all bounded, finitely additive set functions on $\mathcal{P}$ with the total variation norm. For each $\nu \in$ ba, the series $\sum_j \nu(j)$ is absolutely convergent, and $\nu \to \{\nu(j)\}$ defines a projection P from ba onto ℓ^1. Theorem 6 then asserts that if the sequence $\{\mu_i\}$ in ba is Cauchy in the topology $\sigma(ba, m_o)$, then the sequence $\{P\mu_i\}$ is norm convergent in ℓ^1. In particular, the projection P is sequentially continuous with respect to the $\sigma(ba, m_o)$ and norm topologies.

We now give a generalization of Theorem 6 to group-valued measures. Recall that if Σ is a σ-algebra of subsets of a set S, then $\mu : \Sigma \to G$ is strongly additive iff $\lim \mu(E_i) = 0$ for each disjoint sequence $\{E_i\}$ from Σ (§5). If G is complete, a finitely additive set function μ is strongly additive iff the series $\sum \mu(E_i)$ converges for each disjoint sequence $\{E_i\}$ ([34] I.1.18).

Theorem 7. Let G be sequentially complete and let $\mu_i : \Sigma \to G$ be strongly additive. If $\lim_i \mu_i(E) = \mu(E)$ exists for each $E \in \Sigma$, then for each disjoint sequence $\{E_j\}$ from Σ

$\lim_i \sum_{j \in \sigma} \mu_i(E_j) = \sum_{j \in \sigma} \mu(E_j)$ uniformly for $\sigma \subseteq \mathbb{N}$. (In particular, μ is strongly additive.)

Proof: By Theorem 1, it suffices to show that $\lim_i \sum_{j \in \sigma} \mu_i(E_j)$ exists

for each $\sigma \subseteq \mathbb{N}$. First consider the case when $\mu = 0$. We claim that in this case $\lim_{i} \sum_{j\in\sigma} \mu_i(E_j) = 0$ for each σ. If this is not the case, there is a disjoint sequence $\{E_j\}$ such that $\{\sum_{j=1}^{\infty}\mu_i(E_j)\}_i$ doesn't converge to 0. Thus, there exist $\delta > 0$ and a subsequence $\{k_i\}$ such that $|\sum_{j=1}^{\infty}\mu_{k_i}(E_j)| > \delta$. For convenience assume $k_i = i$. Now there exists n_1, such that $|\sum_{j=1}^{n_1}\mu_1(E_j)| > \delta$. There exists m_1 such that $\sum_{j=1}^{n_1}|\mu_i(E_j)| < \delta/2$ for $i \geqslant m_1$. There exists $n_2 > n_1$ such that $|\sum_{j=1}^{n_2}\mu_{m_1}(E_j)| > \delta$. Hence, $|\sum_{j=n_1+1}^{n_2}\mu_{m_1}(E_j)| \geqslant |\sum_{j=1}^{n_2}\mu_{m_1}(E_j)| - \sum_{j=1}^{n_1}|\mu_{m_1}(E_j)| > \delta/2$.

Continuing this construction produces subsequences $\{m_i\}$ and $\{n_i\}$ such that $|\sum_{j=n_i+1}^{n_{i+1}}\mu_{m_i}(E_j)| > \delta/2$. Put $F_i = \bigcup_{j=n_i+1}^{n_{i+1}} E_j$. Then $\{F_j\}$ is a disjoint sequence in Σ with $|\mu_{m_i}(F_i)| > \delta/2$.

Consider the matrix $[z_{ij}] = [\mu_{m_i}(F_j)]$. By Drewnowski's Lemma, $[z_{ij}]$ is a $\mathcal{K}$ matrix. Hence, by the Basic Matrix Theorem, $z_{ii} \to 0$. But $|z_{ii}| = |\mu_{m_i}(F_i)| \geqslant \delta/2$. This contradiction establishes the result in the case when $\mu = 0$.

If $\mu \neq 0$ and $\lim_{i}\sum_{j=1}^{\infty}\mu_i(E_j)$ fails to exist for some disjoint sequence $\{E_j\}$, there exist $\delta > 0$ and a subsequence $\{k_i\}$ such that $|\sum_{j=1}^{\infty}(\mu_{k_{i+1}}(E_j) - \mu_{k_i}(E_j))| > \delta$. Applying the first part to the sequence $\nu_i = \mu_{k_{i+1}} - \mu_{k_i}$ gives the desired contradiction.

We now show that the classical Phillips lemma (Theorem 6) follows as a corollary of Theorem 7.

Proof of Theorem 6: Let $\epsilon > 0$. By Theorem 7,

$|\sum_{j \in \sigma} (\mu_i(j) - \mu(j))| < \epsilon$ for i large. But then

$\sum_{j=1}^{\infty} |\mu_i(j) - \mu(j)| \leqslant 2\epsilon$ for large i ([56] 1.1.2).

Theorem 7 can also be derived from Theorem 1 by employing the Brooks-Jewett result, Theorem 5.6 (see [14]). Most of this section is taken from [14].

9. The Schur Lemma For Bounded Multiplier Convergent Series

In this section we consider bounded multiplier convergent series in a metric linear space and present a version of the Schur lemma for such series. Our result is based on the strengthened version of the Schur lemma for Banach spaces which was given in Corollary 8.3.

Throughout this section X will denote a metric linear space. A series Σx_i in X is said to be <u>bounded</u> <u>multiplier</u> <u>convergent</u> if the series $\Sigma t_i x_i$ is convergent in X for each bounded sequence of scalars $\{t_i\}$. A series Σx_i which is bounded multiplier convergent is clearly subseries convergent (take $t_i = 0$ or 1), but, in general, the converse of this statement is false. In a normed space it is easy to give an example of a series which is subseries convergent but not bounded multiplier convergent by using the normed space in Example 3.5. That is, pick $\{\phi_k\} \in \ell^1$ such that $\phi_k \neq 0$ for each k. Then define a norm on m_o by $\|\{t_i\}\| = \Sigma |t_i \phi_i|$. The series Σe_j is $\|\ \|$-subseries convergent but is not bounded multiplier convergent since, in particular, the series $\Sigma(1/j)e_j$ doesn't converge to an element of m_o. Rolewicz gives an example of a series in an (non-locally convex) F-space which is subseries convergent but not bounded multiplier convergent ([63] III.6.9). In a locally convex F-space a series is subseries convergent iff it is bounded multiplier convergent ([63] III.6.5).

We first establish two preliminary lemmas. The first is an elementary property of the scalar multiplication in a metric linear space which is an immediate corollary of 6.6.

<u>Lemma 1</u>. If $\lim_{j} x_j = 0$ in X, then $\lim_{j} |tx_j| = 0$ uniformly for $|t| \leq 1$.

See also [80] I.2.2 for a proof.

The next lemma is a special case of the general Schur lemma which will be established in Theorem 3.

Lemma 2. Let $x_{ij} \in X$ for $i,j \in \mathbb{N}$ be such that $\sum_j x_{ij}$ is bounded multiplier convergent for each i. If $\lim_i \sum_{j=1}^{\infty} t_j x_{ij} = 0$ for each $\{t_j\} \in \ell^{\infty}$, then $\lim_i \sum_{j=1}^{\infty} t_j x_{ij} = 0$ uniformly for $\|\{t_j\}\| \leq 1$.

Proof: If the conclusion fails to hold, we may assume (by passing to a subsequence if necessary) that there is a $\delta > 0$ such that $\sup\{ |\sum_{j=1}^{\infty} t_j x_{ij}| : |t_j| \leq 1 \} > \delta$ for each i. Set $i_1 = 1$. Then there exists $\alpha_1 = \{t_{1j}\} \in \ell^{\infty}$ such that $\|\alpha_1\| \leq 1$ and $|\sum_{j=1}^{\infty} t_{1j} x_{i_1 j}| > \delta$.

There exists M_1 such that $|\sum_{j=1}^{M_1} t_{1j} x_{i_1 j}| > \delta$. Since $\lim_i x_{ij} = 0$ for each j, by the observation in Lemma 1 above there exists $i_2 > i_1$ such that $i \geq i_2$ implies $\sum_{j=1}^{M_1} |t_j x_{ij}| < \delta/2$ for $|t_j| \leq 1$.

Now there exists $\alpha_2 = \{t_{2j}\} \in \ell^{\infty}$ such that $\|\alpha_2\| \leq 1$ and $|\sum_{j=1}^{M_2} t_{2j} x_{i_2 j}| > \delta$. There exists $M_2 > M_1$, such that $|\sum_{j=1}^{M_2} t_{2j} x_{i_2 j}| > \delta$.

Note $|\sum_{j=M_1+1}^{M_2} t_{2j} x_{i_2 j}| \geq |\sum_{j=1}^{M_2} t_{2j} x_{i_2 j}| - \sum_{j=1}^{M_1} |t_{2j} x_{i_2 j}| > \delta/2$.

Continuing this construction inductively gives subsequences $\{i_k\}$ and $\{M_k\}$ of positive integers such that

(1)
$$\left| \sum_{j=M_{k-1}+1}^{M_k} t_{kj} x_{i_k j} \right| > \delta/2 \quad \text{for all} \quad k,$$

where $M_0 = 0$.

Now consider the matrix $[z_{kp}] = [\sum_{j=M_{p-1}+1}^{M_p} t_{pi} x_{i_p j}]$. Since $\lim_j x_{ji} = 0$ for each i, $\lim_k z_{kp} = 0$ for each p since $|t_{pi}| \leqslant 1$. We claim that $[z_{kp}]$ is a $\mathcal{K}$ matrix. Let $\sigma \subseteq \mathbb{N}$. Define a sequence $\{s_j\} \in \ell^\infty$ by $s_j = t_{pj}$ if $M_{p-1} + 1 \leqslant i \leqslant M_p$ and $j \in \sigma$ and $s_j = 0$ otherwise. Then $\sum_{p \in \sigma} z_{kp} = \sum_{j=1}^\infty s_j x_{i_k j}$ converges to 0 as $k \to \infty$ by hypothesis. Thus, the matrix $[z_{kp}]$ is a $\mathcal{K}$ matrix and the Basic Matrix Theorem implies that $\lim_k z_{kk} = 0$. But this contradicts (1).

We now present our Schur-type lemma for bounded multiplier convergent series. Note that conclusion (ii) in the theorem below is just the conclusion of the Schur lemma for B-spaces given in 8.3.

Theorem 3. Let $x_{ij} \in X$ for $i, j \in \mathbb{N}$ be such that $\sum_j x_{ij}$ is bounded multiplier convergent for each i. Assume that $\lim_i \sum_{j=1}^\infty t_j x_{ij}$ exists for each $\{t_j\} \in \ell^\infty$. If $\lim_i x_{ij} = x_j$, then

(i) the series $\sum x_j$ is bounded multiplier convergent and

(ii) $\lim_i \sum_{j=1}^\infty t_j x_{ij} = \sum_{j=1}^\infty t_j x_j$ uniformly for $||\{t_j\}|| \leqslant 1$.

Proof: First we claim that the sequence $\{\sum_{j=1}^\infty t_j x_{ij}\}$ satisfies a Cauchy condition uniformly for $||\{t_j\}|| \leqslant 1$. If this is not the case, there exist a subsequence $\{i_k\}$ and a $\delta > 0$ such that

(2) $\qquad \sup\{ | \sum\limits_{j=1}^{\infty} t_j(x_{i_{k+1}j} - x_{i_k j})| : ||\{t_j\}|| \leq 1 \} > \delta.$

Consider the series $\sum\limits_{j}(x_{i_{k+1}j} - x_{i_k j})$. This sequence of series satis-

fies the hypothesis of Lemma 2 so that $\lim\limits_{k} \sum\limits_{j=1}^{\infty} t_j(x_{i_{k+1}j} - x_{i_k j}) = 0$

uniformly for $||\{t_j\}|| \leq 1$. This contradicts (2) and establishes

the claim.

We now establish (i) and also

(3) $\qquad \lim\limits_{i} \sum\limits_{j=1}^{\infty} t_j x_{ij} = \sum\limits_{j=1}^{\infty} t_j x_j$ for each $\{t_j\} \in \ell^{\infty}.$

Let $\epsilon > 0$. By what has been established above, there exists an

N such that $i,k \geq N$ implies $| \sum\limits_{j \in \sigma} t_j(x_{ij} - x_{kj})| < \epsilon$ for

$\sigma \subseteq N$. Hence, for every M and $i \geq N$ $| \sum\limits_{j=1}^{M} t_j(x_{ij} - x_j)| \leq \epsilon.$

Thus, for $i \geq N$,

(4) $\qquad | \sum\limits_{j=1}^{M} t_j x_j - \sum\limits_{j=1}^{\infty} t_j x_{ij}| \leq | \sum\limits_{j=1}^{M} t_j(x_j - x_{ij})|$

$\qquad + | \sum\limits_{j=M+1}^{\infty} t_j(x_{Nj} - x_{ij})| + | \sum\limits_{j=M+1}^{\infty} t_j x_{Nj}| < 2\epsilon + | \sum\limits_{j=M+1}^{\infty} t_j x_{Nj}|.$

The last term on the right hand side of (4) goes to 0 as $M \to \infty$

for N fixed. Condition (i) and (3) follow from this estimate.

Condition (ii) follows by applying Lemma 2 to the series

$\sum\limits_{j}(x_{ij} - x_j).$

Theorem 3 has as a corollary a generalization of another result

of Schur on summability which was discussed previously in 5.15, 8.4

and 8.5. Let $A = [a_{ij}]$ be a real matrix such that the sequence

$\{\sum_{j=1}^{\infty} t_j a_{ij}\}$ is convergent for each $\{t_j\} \in \ell^{\infty}$, i. e., A is of class

(ℓ^{∞}, c). The Schur summability result then asserts that the series

$\sum_j |a_{ij}|$ converge uniformly in i (Corollary 5.15 or [50] 7.6).

This condition (for real series) clearly implies that the series

$\sum_j t_j a_{ij}$ are uniformly convergent for $||\{t_j\}|| \leqslant 1$ and $i \in \mathbf{N}$.

Using Theorem 3 we obtain the analogous result for bounded multi-plier convergent series in X.

Corollary 4. Let x_{ij} satisfy the hypothesis of Theorem 3. Then

the series $\sum_j t_j x_{ij}$ converge uniformly for $||\{t_j\}|| \leqslant 1$ and $i \in \mathbf{N}$.

Proof: First note that if the series $\sum_j x_j$ is bounded multiplier

convergent, then the series $\sum_j t_j x_j$ converge uniformly for

$||\{t_j\}|| \leqslant 1$. (This follows from Theorem 3 but is also easily

checked directly.)

Let $\epsilon > 0$. By Theorem 3, there exists N such that $i \geqslant N$

implies $|\sum_{j=1}^{\infty} t_j (x_{ij} - x_j)| < \epsilon$ for $||\{t_j\}|| \leqslant 1$. By the observation

above, there exists M such that $m \geqslant M$ implies

$|\sum_{j=m}^{\infty} t_j x_{ij}| < \epsilon$, $|\sum_{j=m}^{\infty} t_j x_j| < \epsilon$ for $1 \leqslant i \leqslant N$, $||\{t_j\}|| \leqslant 1$.

Hence, $m \geqslant M$ implies

$$|\sum_{j=m}^{\infty} t_j x_{ij}| \leqslant |\sum_{j=m}^{\infty} t_j (x_{ij} - x_j)| + |\sum_{j=m}^{\infty} t_j x_j| < 2\epsilon$$

for $i \geqslant N$, $||\{t_j\}|| \leqslant 1$, and the result follows.

Corollary 4 can be used to obtain a characterization of

matrices of class $(\ell^{\infty}, c(X))$ when X is a metric linear space (see

8.5). Recall the matrix $A = [x_{ij}]$ is of class $(\ell^\infty, c(X))((m_o, c(X)))$ if the sequence $\{\sum_j t_j x_{ij}\}$ is convergent in X for each sequence $\{t_j\} \in \ell^\infty$ ($\{t_j\} \in m_o$). From Corollary 4 we have the following vector version of the Schur summability theorem for metric linear spaces.

<u>Corollary 5</u>. Let the matrix $A = [x_{ij}]$ be such that the rows are bounded multiplier covergent. The following are equivalent:

(a) $A \in (\ell^\infty, c(X))$

(b) (i) the series $\sum_j t_j x_{ij}$ converge uniformly for $i \in N$

   and $||\{t_j\}|| \leqslant 1$

   (ii) $\lim_i x_{ij} = x_j$ exists for each j.

(c) $\lim_i \sum_{j=1}^\infty t_j x_{ij}$ exists uniformly for $||\{t_j\}|| \leqslant 1$.

<u>Proof</u>: That (a) implies (b) (i) follows from Corollary 4. That (a) implies (b) (ii) follows by setting $\{t_j\} = e_j$.

   Suppose (b) holds. Let $\epsilon > 0$. By (i) there exists N such that $|\sum_{j=N}^\infty t_j x_{ij}| < \epsilon$ for $i \in N$ and $|t_j| \leqslant 1$. Then (ii) implies that $|\sum_{j=N}^\infty t_j x_j| \leqslant \epsilon$ for $|t_j| \leqslant 1$. By (ii) and Lemma 1 there exists $M > 0$ such that $\sum_{j=1}^{N-1} |t_j(x_{ij}-x_j)| < \epsilon$ for $|t_j| \leqslant 1$ and $i \geqslant M$.

Thus for $i \geqslant M$ and $|t_j| \leqslant 1$,

$$|\sum_{j=1}^\infty t_j(x_{ij}-x_j)| \leqslant \sum_{j=1}^{N-1} |t_j(x_{ij}-x_j)| + |\sum_{j=N}^\infty t_j x_{ij}| + |\sum_{j=N}^\infty t_j x_j| < 3\epsilon,$$

and (c) holds.

That (c) implies (a) is clear.

This result along with the conclusion of Theorem 8.5 suggests the following question. If the matrix A is of class $(m_0, c(X))$, is A also of class $(\ell^\infty, c(X))$? If X is a B-space, this is the case (Theorem 8.5). We show that this is not the case for metric linear spaces. Of course, if the rows of the matrix are only subseries convergent and not bounded multiplier convergent, the matrix could not be of class $(\ell^\infty, c(X))$, but we give an example of a matrix whose rows are bounded multiplier convergent and which belongs to $(m_0, c(X))$ but not $(\ell^\infty, c(X))$.

Let X be a metric linear space containing a series Σx_j which is subseries convergent but not bounded multiplier convergent. (For an example in a normed space, see the example given in the introduction to this section; for an example in a complete metric linear space see [63] III.6.9.) Set

$x_{ij} = x_j$ if $1 \leqslant j \leqslant i$ and $x_{ij} = 0$ if $j > i$. Then $\sum_j x_{ij}$

is subseries convergent, and for each i, the sequence $\{x_{ij}\}_{j=1}^\infty$ lies in a finite dimensional subspace of X so that $\sum_j x_{ij}$ is

bounded multiplier convergent. By the subseries convergence of Σx_j, we have $\lim_i \sum_{j \in \sigma} x_{ij} = \sum_{j \in \sigma} x_j$ for each $\sigma \subseteq \mathbb{N}$; that is,

$A = [x_{ij}]$ is of class $(m_0, c(X))$. But, since $\sum_j x_j$ is not

bounded multiplier convergent, condition (i) of Theorem 3 fails to hold; that is, A is not of class $(\ell^\infty, c(X))$.

Actually, Theorem 8.5 and its proof shows that A belongs to $(m_0, c(X))$ iff condition (c) or (d) of Theorem 8.5 holds. This observation along with Corollary 5 give characterizations of the classes $(\ell^\infty, c(X))$ and $(m_0, c(X))$ in the case when X is a metric linear space.

The example given above also shows that the hypothesis in

Theorem 3 cannot be replaced by the weaker hypothesis that $\lim\limits_{i} \sum\limits_{j \in \sigma} x_{ij}$

exists for each $\sigma \subseteq \mathbb{N}$.

Another interesting type of series generated by multiplying a given series by certain sequences of scalars is also discussed by Rolewicz in [63]. Rolewicz calls a series Σx_j in a metric linear space X a C-series if the series $\Sigma t_i x_i$ converges in X for each scalar sequence $\{t_i\} \in c_0$. (This is a slight departure from the terminology of Rolewicz ([63] III.8).) These series have been studied in detail in the case of normed spaces and it is known that a Banach space X has the property that every C-series is (subseries) convergent iff X contains no subspace (topologically) isomorphic to c_0 ([21]) (The series Σe_i in c_0 is C-convergent but not convergent.) A natural question that arises in the light of Theorem 3 is whether the analogue of Theorem 3 is valid for C-convergent series. We first note that the analogue of Theorem 3 (i) is indeed valid for C-convergent series.

<u>Proposition 6</u>. Let $x_{ij} \in X$ be such that $\sum\limits_{j} x_{ij}$ is C-convergent

for each i. Assume that $\lim\limits_{i} \sum\limits_{j=1}^{\infty} t_j x_{ij}$ exists for each $\{t_j\} \in c_0$.

If $x_j = \lim\limits_{i} x_{ij}$, then the series Σx_j is C-convergent.

<u>Proof</u>: Let $\{t_j\} \in c_0$. Note that for each $\{s_j\} \in \ell^\infty$, the sequence $\{s_j t_j\} \in c_0$. Therefore, Theorem 3 can be applied to the series $\sum\limits_{j} t_j x_{ij}$, and condition (i) implies that the series $\Sigma t_j x_j$ is convergent.

Whereas the analogue of Theorem 3 (i) does hold for C-convergent series, it is easy to see that the analogue of Theorem 3 (ii) does not hold for such series. For an example, let e_k be the unit vector in c_o. Set $e_{ij} = e_j$ for $1 \leqslant j \leqslant i$ and $e_{ij} = 0$ for $j > i$. Then each series $\underset{j}{\Sigma} e_{ij}$ is C-convergent. For $\{t_j\} \in c_o$,

$$\lim_i \sum_{j=1}^{\infty} t_j e_{ij} = \lim_i \sum_{j=1}^{i} t_j e_j = \sum_{j=1}^{\infty} t_j e_j,$$ but the convergence is not uniform

for $||\{t_j\}|| \leqslant 1$. Note also that the analogue of the conclusion in Corollary 8.4 is false for this particular example.

Much of the material from this section is contained in [73].

10. Imbedding c_0 and ℓ^∞

In this section we consider imbedding the classical sequence spaces c_0 and ℓ^∞ in a given Banach space X. In Theorem 3 we establish a general result which gives a sufficient condition for a Banach space to contain a subspace isomorphic to c_0. This general result is then employed to give the Bessaga-Pelczynski characterization of B-spaces which contain subspaces isomorphic to c_0, a result of Diestel on vector measures and a result of Pelczynski on unconditionally converging operators. In Theorem 7 we give a sufficient condition for a B-space to contain a subspace isomorphic to ℓ^∞. As a corollary of Theorem 7 we obtain the Diestel-Faires characterization of B-spaces containing a subspace isomorphic to ℓ^∞ as well as a result of Rosenthal on bounded linear operators on ℓ^∞. The results and style of proof are very similar to those of section I.4 of Diestel and Uhl ([29]). Whereas Diestel and Uhl employ Rosenthal's lemma, we use the Basic Matrix Lemma.

Throughout this section X will denote a Banach space. A series Σx_i in X is said to be weakly unconditionally Cauchy (w.u.c.) if $\Sigma |< x',x_i >| < \infty$ for each $x' \in X'$ (such series are sometimes called weakly unconditionally convergent ([21])). Such series may not be convergent in X; for example, consider the series Σe_i in c_0. The following result gives several characterizations of w.u.c. series which will be needed later.

Proposition 1. Let $\{x_i\} \subseteq X$. The following are equivalent:
  (i) Σx_i is w.u.c.
  (ii) $\{\Sigma_{i \in \sigma} x_i : \sigma \subseteq \mathbb{N} \text{ finite}\}$ is (norm) bounded
  (iii) for each $\{t_i\} \in c_0$, $\Sigma t_i x_i$ converges

(iv) $\sup\{ \sum\limits_{i=1}^{\infty} | < x', x_i > | \; : \; ||x'|| \leqslant 1 \} = M < \infty$ and

$$|| \sum\limits_{i=1}^{\infty} t_i x_i || \leqslant M||\{t_i\}|| \text{ for each } \{t_i\} \in c_o.$$

(v) $|| \sum\limits_{i=1}^{n} t_i x_i || \leqslant M ||\{t_i\}||$ for each $\{t_i\} \in c_o$ and $n \in \mathbb{N}$.

Proof: Assume (i) and let $\mathcal{J}$ be the finite subsets of $\mathbb{N}$. Then for

$$x' \in X', | < x, \sum\limits_{i \in \sigma} x_i > | \leqslant \sum\limits_{i=1}^{\infty} | < x', x_i > |$$

for each $\sigma \in \mathcal{J}$. Hence, the set $\{ \sum\limits_{i \in \sigma} x_i : \sigma \in \mathcal{J} \}$ is weakly bounded

and, thus, norm bounded. That is, (ii) holds.

Assume (ii). Let $M > 0$ be such that $|| \sum\limits_{i \in \sigma} x_i || \leqslant M$ for

$\sigma \in \mathcal{J}$. If $||x'|| \leqslant 1$, $| \sum\limits_{i \in \sigma} < x', x_i > | \leqslant M$ for $\sigma \in \mathcal{J}$ so that

$\sum\limits_{i \in \sigma} | < x', x_i > | \leqslant 2M$ ([56] 1.1.2). If $\{t_i\} \in c_o$ and

$n > m$, $|| \sum\limits_{i=m}^{n} t_i x_i || = \sup\{ | \sum\limits_{i=m}^{n} t_i < x', x_i > | \; : \; ||x'|| \leqslant 1 \}$

$\leqslant 2M \sup\{ |t_i| : m \leqslant i \leqslant n \}$ and   (iii) holds.

Assume (iii). For $\{t_i\} \in c_o$, we have

(1) $|| \sum\limits_{i=1}^{\infty} t_i x_i || = \sup\{ \sum\limits_{i=1}^{\infty} t_i < x', x_i > : ||x'|| \leqslant 1 \} < \infty$.

Thus, for each $x' \in X'$, $||x'|| \leqslant 1$, the sequence $\{ < x', x_i > \}$ is

in ℓ^1, and (1) implies that the set

$$B = \{ \{ < x', x_i > \} \in \ell^1 : ||x'|| \leqslant 1 \}$$

is weak*-bounded in ℓ^1. Hence, B is norm bounded in ℓ^1, i.e.,

$$\sup\{ \sum_{i=1}^{\infty} |<x',x_i>| : ||x'|| \leqslant 1\} = M < \infty.$$

For $\{t_i\} \in c_o$, (1) implies that $||\sum_{i=1}^{\infty} t_i x_i|| \leqslant M \, ||\{t_i\}||$, or (iv) holds.

That (iv) implies (i) and (v) is clear.

Thus, (i) - (iv) are equivalent and since (iv) implies (v), it suffices to show that (v) implies (iii).

Assume that (v) holds. Define the linear operator T on the linear subspace c_{oo} of c_o consisting of the finitely non-zero sequences by $T\{t_i\} = \sum_{i=1}^{\infty} t_i x_i$. Then (v) implies that T is a bounded linear operator on c_{oo} and, therefore, has a bounded linear extension, still denoted by T, to c_o with norm less than or equal to M. Thus, if $\{t_i\} \in c_o$, we have $T\{t_i\} = \sum_{i=1}^{\infty} t_i x_i$, and (iii) holds.

These properties of w.u.c. series are well-known and most of them are given in [21] 5.2.

We next derive the basic matrix result which will be employed to obtain the main results of this section. This result is based on the basic matrix Lemma 2.1 and is a simple consequence of Lemma 2.1.

<u>Lemma 2</u>. Let $x_{ij} \in X$ be such that $\lim_i x_{ij} = 0$ for each j and $\lim_j x_{ij} = 0$ for each i. Given $\epsilon > 0$ there exists a subsequence $\{m_i\}$ such that $\sum_{i=1}^{\infty} \sum_{\substack{j\neq i \\ j=1}}^{\infty} ||x_{m_i m_j}|| < \epsilon$.

<u>Proof</u>: Pick $\epsilon_{ij} > 0$ such that $\sum_{i,j} \epsilon_{ij} < \epsilon$ (for example, $\epsilon_{ij} = \epsilon/2^{i+j+1}$). Let $\{m_i\}$ be the subsequence given by the matrix

Lemma 2.1 (applied to $||x_{ij}||$). Then $||x_{m_i m_j}|| < \epsilon_{ij}$ for $i \neq j$ and the result follows easily.

Using Lemma 2 we now establish our main result concerning the imbedding of c_o in X.

Theorem 3. Suppose that X contains a w.u.c. series Σx_i which is such that $||x_i|| \geqslant \delta > 0$ for each i. Then there exists a subsequence $\{m_i\}$ such that for any subsequence $\{n_i\}$ of $\{m_i\}$ $T\{t_i\} = \overset{\infty}{\underset{i=1}{\Sigma}} t_i x_{n_i}$ defines a topological isomorphism T of c_o into X.

Proof: By replacing X by the closed subspace generated by the $\{x_i\}$, we may assume that X is separable. For each i pick $x_i' \in X'$ such that $||x_i'|| = 1$ and $<x_i',x_i> = ||x_i||$. By the Banach-Alaoglu Theorem $\{x_i'\}$ has a subsequence which converges weak* to an element $x' \in X'$. To avoid cumbersome notation later, assume that $x_i' \to x'$ weak*. Now $|<x_i' - x',x_i>| \geqslant \delta - |<x', x_i>| \geqslant \delta/2$ for large i since $\lim <x',x_i> = 0$. Again to avoid cumbersome notation, assume that $|<x_i' - x', x_i>| > \delta/2$ for all i.

The matrix $[<x_i' - x',x_j>]$ satisfies the condition of the Matrix Lemma 2. Let $\{m_i\}$ be the subsequence given by the conclusion of lemma 2 with $\epsilon = \delta/4$.

Now define a bounded linear operator $T : c_o \to X$ by $T\{t_i\} = \overset{\infty}{\underset{i=1}{\Sigma}} t_i x_{m_i}$. (Note this map is well-defined and is continuous by Proposition 1.) If $z_i' = x_{m_i}' - x'$, then, using the conclusion of Lemma 2,

$$2||T\{t_i\}|| \geqslant | <z_i', T\{t_j\}> | \geqslant |t_i <z_i',x_{m_i}>| - \sum_{j\neq i} |t_j| |<z_i',x_{m_j}>|$$

$\geqslant |t_i|\delta/2 - ||\{t_j\}||\delta/4$. Taking the supremum over i gives $2||T\{t_j\}|| \geqslant ||\{t_j\}||\delta/4$ which implies that T has a bounded inverse.

The same computation holds for any subsequence $\{n_i\}$ of $\{m_i\}$.

This result was derived in [69] by using a form of the Antosik-Mikusinski Diagonal Theorem. (See also [13].) The proof given here is somewhat simpler and is based on Lemma 2 which is a more elementary result than the Diagonal Theorem.

We now give several applications of Theorem 3. First we derive a classic result of Bessaga and Pelczynski on w.u.c. series ([21]).

Corollary 4. The B-space X is such that every w.u.c. series in X is subseries convergent (norm) iff X contains no subspace isomorphic to c_o.

Proof: Suppose X contains a series Σx_i which is w.u.c. but not subseries convergent. Then there is a subseries Σx_{n_i} which does not converge. Hence, there is a $\delta > 0$ and an increasing sequence $\{p_j\}$ in $\mathbb{N}$ such that $||z_j|| > \delta$ for each j, where

$z_j = \sum_{i=p_j+1}^{p_{j+1}} x_{n_i}$. By Proposition 1 (ii), the series Σz_j is w.u.c.

and satisfies the hypothesis of Theorem 3. By Theorem 3, X contains a subspace isomorphic to c_o.

The other implication is obvious since c_o contains a series which is w.u.c. but not subseries convergent (namely, Σe_j).

Bessaga and Pelczynski derive Corollary 4 from results on basic sequences in B-spaces ([21]); Diestel and Uhl give a proof of Corollary 4 based on a result of Rosenthal ([34] I.4.5). The methods employed in the proof given above should be contrasted with those of

Diestel and Uhl ([34]) where the Rosenthal lemma is used.

We next derive a result of Diestel on vector measures from Theorem 3 ([34] I.4.2).

Corollary 5. Let A be an algebra of subsets of a set S. If $m : A \to X$ is a bounded, finitely additive set function which is not strongly additive, then X contains a subspace isomorphic to c_o.

Proof: If m is not strongly additive, there is a disjoint sequence $\{A_i\}$ in A and a $\delta > 0$ such that $||m(A_i)|| > \delta$ for each i. (See chapters 1 or 4.) For $x' \in X'$, the A scalar set function $x'm$ is bounded and finitely additive and, therefore, has bounded variation. Hence, $\Sigma |<x',m(A_j)>| < \infty$. That is, the series $\Sigma m(A_j)$ is w.u.c. Theorem 3 now gives the result.

A proof of Corollary 5 is given by Diestel and Uhl in [34] and is based on Rosenthal's Lemma.

Finally we use Theorem 3 to derive a result of Pelczynski on unconditionally converging operators. A bounded linear operator T from a B-space X into a B-space Y is an unconditionally converging operator if T carries w.u.c. series in X into subseries convergent series in Y ([58]). A weakly compact operator is unconditionally converging and in certain B-spaces the converse is also true ([58]). The identity operator on ℓ^1 gives an example of an operator which is unconditionally converging but not weakly compact. We have the following interesting result of Pelczynski on unconditionally converging operators ([58]).

Corollary 6. Let $T : X \to Y$ be a bounded linear operator which is not unconditionally converging. Then there exist isomorphisms $I_1 : c_o \to X$ and $I_2 : c_o \to Y$ such that $TI_1 = I_2$ (i.e., T has a bounded inverse on a subspace isomorphic to c_o).

Proof: By hypothesis there exists a w.u.c. series Σx_i in X such that $\Sigma T x_i$ is w.u.c. but not subseries convergent. Since $\Sigma T x_i$ contains a subseries which is not convergent, we may as well assume that $\Sigma T x_i$ is not convergent. Thus, there exist $\delta > 0$ and a subsequence $\{p_i\}$ such that $||z_j|| > \delta$, where $z_j = T u_j$ and

$u_j = \sum\limits_{i=p_j+1}^{p_{j+1}} x_i$. By Proposition 1 (ii), the series $\Sigma T u_j$ is w.u.c.

Since $||x|| \geqslant || Tx || / ||T||$ for each $x \in X$, the series Σu_j is w.u.c. in Y by Proposition 1 (ii) and, moreover, $||u_j|| > \delta / ||T||$. Applying Theorem 3 to the series Σu_j and $\Sigma T u_j$ there is a subsequence $\{m_i\}$ such that $I_1\{t_j\} = \Sigma t_j u_{m_j}$ and

$I_2\{t_j\} = \Sigma t_j T u_{m_j}$ define isomorphisms I_1 and I_2 from c_o into X and Y, respectively. Evidently $T I_1 = I_2$.

Pelczynski derives Corollary 6 by using a deep theorem of [21] on the existence of basic sequences ([58]). Corollary 6 is established in [69] by employing the Antosik-Mikusinski Diagonal Theorem. The proof above is of a much more elementary character.

The converse of Corollary 6 holds and furnishes an interesting characterization of unconditionally converging operators ([43]).

We now consider the problem of imbedding ℓ^∞ in X.

In what follows, let $\mathcal{P}$ be the power set of N. If $J \subseteq N$, $\ell^\infty(J)$ will denote the subspace of ℓ^∞ which consists of those sequences which vanish outside of J. We establish the analogue of Theorem 3 for measures.

Theorem 7. Let $\mu : \mathcal{P} \to X$ be bounded and finitely additive. If $\{\mu(j)\}$ doesn't converge to 0, then there exists a subsequence

$\{n_i\}$ such that for any subsequence $\{p_i\}$ of $\{n_i\}$, $T\phi = \int_J \phi d\mu$ defines a topological isomorphism $T : \ell^\infty(J) \to X$, where $J = \{p_j : j \in \mathbb{N}\}$.

Proof: We may assume (by passing to a subsequence if necessary) that there exists a $\delta > 0$ such that $||\mu(j)|| > \delta$ for each j. Pick $x_j' \in X'$ such that $||x_j'|| = 1$ and $\langle x_j', \mu(j)\rangle = ||\mu(j)||$. Let X_0 be the closed subspace spanned by $\{\mu(j) : j \in \mathbb{N}\}$. Let z_j' be x_j' restricted to the subspace X_0. Since X_0 is separable, $\{z_j'\}$ is $\sigma(X_0', X_0)$ relatively sequentially compact and, therefore, has a subsequence which is $\sigma(X_0', X_0)$ convergent to an element $z' \in X_0'$ with $||z'|| \leqslant 1$. For convenience of notation, assume that $\{z_j'\}$ is $\sigma(X_0', X_0)$ convergent to z'. Extend z' to an element $x' \in X'$ with $||x'|| \leqslant 1$. Thus, we have

(2) $\qquad \lim_j \langle x_j' - x', \mu(i)\rangle = 0$ for each i.

Consider the matrix $[y_{ij}] = [\langle x_i' - x', \mu(j)\rangle]$. Now $|y_{ii}| > \delta - |\langle x', \mu(i)\rangle| > \delta/2$ for large i since $x'\mu$ is strongly additive so, again for convenience of notation, assume that

(3) $\qquad |y_{ii}| > \delta/2$ for all i.

By (2) and the strong additivity of each $(x_i' - x')\mu$, we have $\lim_i y_{ij} = 0$ for each j and $\lim_j y_{ij} = 0$ for each i.

Therefore, we may apply Lemma 2 and obtain a subsequence $\{m_i\}$ such that

(4) $\qquad \sum_{i=1}^\infty \sum_{j \neq i} |y_{m_i m_j}| < \delta/4$.

Since each $(x_i' - x')\mu$ is strongly additive, Drewnowski's Lemma implies that $\{m_i\}$ has a subsequence $\{n_i\}$ such that each $(x_{m_i}' - x')\mu$ is countably additive on the σ-algebra generated by

$\{n_i\}$. Put $J = \{n_i : i \in \mathbb{N}\}$. Define a bounded linear operator $T : \ell^\infty(J) \to X$ by $T\phi = \int_J \phi d\mu$.

From the countable additivity, (3) and (4), we have

$$(5) \qquad 2||T\phi|| \geqslant |\langle x'_{n_i} - x', \int_J \phi d\mu\rangle| = |\int_J \phi d(x'_{n_i} - x')\mu| =$$

$$|\sum_{j=1}^{\infty} t_{n_j} \langle x'_{n_i} - x', \mu(n_j)\rangle| \geqslant |t_{n_i}| |y_{n_i n_i}| -$$

$$\sum_{j \neq i} |t_{n_j}| |y_{n_i n_j}| > |t_{n_i}| \delta/2 - ||\phi|| \delta/4,$$

where $\phi = \{t_j\}$. Taking the supremum over i in (5) implies $||T\phi|| \geqslant ||\phi|| \delta/8$, or T has a bounded inverse.

The same calculation holds for any subsequence of $\{n_i\}$.

This result is established in [13] by means of another matrix type result. (See also [72].)

Theorem 7 has an immediate corollary the following result of Diestel and Faires ([33] I.4.2, [34]).

Corollary 8. Let Σ be a σ-algebra of subsets of a set S and let $m : \Sigma \to X$ be bounded, finitely additive but not strongly additive. Then X contains a subspace isomorphic to ℓ^∞.

Proof: If m is not strongly additive, there are a disjoint sequence $\{E_j\} \subseteq \Sigma$ and a $\delta \geqslant 0$ such that $||m(E_j)|| \geqslant \delta$. Define $\mu : P \to X$ by $\mu(A) = m(\underset{j \in A}{\cup} E_j)$. Then μ satisfies the conditions of Theorem 7 and the result is immediate.

Note that the converse of Corollary 8 also holds and, therefore, gives an interesting characterization of B-spaces containing subspaces isomorphic to ℓ^∞. (The set function $\mu : P \to \ell^\infty$ defined by $\mu(\sigma) = \underset{i \in \sigma}{\sum} e_i$ is bounded, finitely additive but not strongly additive.)

Corollary 8 has a number of interesting applications (see [33] I.4). We present two such applications. First we consider the following result due to Bessaga and Pelczynski ([21]).

Corollary 9. If X' contains no subspace isomorphic to ℓ^∞, then X' contains no subspace isomorphic to c_o.

Proof: Let $\Sigma x_i'$ be a w.u.c. series in X'. Then, in particular, $\Sigma|\langle x_i',x\rangle| < \infty$ for each $x \in X$. Thus, the series $\Sigma x_i'$ is weak* subseries convergent. Now define a bounded, finitely additive set function $\mu : \mathcal{P} \to X'$ by $\mu(\sigma) = \sum\limits_{i\in\sigma} x_i'$. By Corollary 8, μ is strongly additive. Hence, for any subsequence $\{m_i\}$, $\lim \|\mu(m_i)\| = \lim \|x_{m_i}'\| = 0$. Theorem 7.1 implies that the series $\Sigma x_i'$ is norm subseries convergent. Corollary 4 gives the result.

As a second application, we establish the result of Diestel and Faires on weak* subseries convergent series which was referred to earlier in sections 3 and 7.

Corollary 10. Let X' contain no subspace isomorphic to ℓ^∞. If $\Sigma x_i'$ is weak* subseries convergent in X', then $\Sigma x_i'$ is norm subseries convergent.

Proof: Define a bounded, finitely additive set function $\mu : \mathcal{P} \to X'$ by $\mu(\sigma) = \sum\limits_{i\in\sigma} x_i'$. By Corollary 8 μ is strongly additive. Hence, for any subsequence $\{m_i\}$, $\lim \mu(m_i) = \lim x_{m_i}' = 0$ in norm. Theorem 7.1 implies that the series $\Sigma x_i'$ is norm subseries convergent.

The converse of Corollary 10 also holds and gives an interesting characterization of dual spaces containing copies of ℓ^∞ (see [33]

Corollary 1.2 for the proof).

Finally, we obtain as a corollary of Theorem 7 the following result of Rosenthal ([65]).

<u>Corollary 11</u>. (Rosenthal) Let $T : \ell^\infty \to X$ be bounded and linear. If there is an infinite $I \subseteq \mathbb{N}$ such that T restricted to $c_o(I)$ is an isomorphism, then there is an infinite $J \subseteq I$ such that T restricted to $\ell^\infty(J)$ is an isomorphism.

<u>Proof</u>: Define $\mu : \mathcal{P} \to X$ by $\mu(A) = T(C_A)$, where C_A denotes the characteristic function of A. Then μ is bounded, finitely additive and $Tf = \int_{\mathbb{N}} f d\mu$. By hypotheses there is a $\delta > 0$ such that

$$\|\mu(i)\| = \|TC_{\{i\}}\| \geqslant \delta \quad \text{for} \quad i \in I.$$ Thus, Theorem 7 gives the result.

Rosenthal actually obtains a more general result than Corollary 11 in Proposition 1.2 of [65]. For the countable case of his result which is given in Corollary 11 our methods are much simpler.

REFERENCES

1. A. Alexiewicz, On Sequences of Operations II, Studia Math., 11 (1950), 200-236.

2. P. Antosik, On the Mikusinski Diagonal Theorem, Bull. Acad. Polon. Sci., 19 (1971), 305-310.

3. P. Antosik, A Generalization of the Diagonal Theorem, Bull. Acad. Polon. Sci., 20 (1972), 373-377.

4. P. Antosik, Mappings from L-groups into Topological Groups I, Bull. Acad. Polon. Sci., 21 (1973), 145-153.

5. P. Antosik, Mappings from L-groups into Topological Groups II, Bull. Acad. Polon. Sci., 21 (1973), 155-160.

6. P. Antosik, Sur les suites d'applictions, C. R. Acad. Sci., Paris, 287 (1978), 75-77.

7. P. Antosik, Mappings from convergence groups into quasi-normed groups, Serdica, 3 (1977), 176-179.

8. P. Antosik, Permutationally convergent matrices, Serdica, 3 (1977), 198-205.

9. P. Antosik, A Diagonal Theorem for Nonnegative Matrices and Equicontinuous Sequences of Mappings, Bull. Acad. Polon. Sci., 24 (1976), 955-959.

10. P. Antosik, On uniform boundedness of families of mappings, Proceedings of the Conference on Convergence Spaces, Szcryrk, 1979, 2-16.

11. P. Antosik, On convergences in sequence spaces, Proceedings of the Conference on Generalized Functions and their Applications, Moscow, 1980.

12. P. Antosik, J. Mikusinski, and R. Sikorski, Theory of Distributions, PWN-Warsaw, 1973.

13. P. Antosik and C. Swartz, A Theoem on Matrices and its Applications in Functional Analysis, Studia Math., 77 (1984), 197-205.

14. P. Antosik and C. Swartz, The Nikodym Boundedness Theorem and the Uniform Boundedness Principle, Proceedings of the Workshop on the Measure Theory and its Applications, Sherbrooke, Quebec, 1982, Springer Lecture Notes on Mathematics, 1033, 36-42.

15. P. Antosik and C. Swartz, The Schur and Phillips Lemmas for Topological Groups, J. Math. Anal. Appl., 98 (1984), 179-187.

16. P. Antosik and C. Swartz, Matrix Methods in Analysis, preprint.

17. P. Antosik and C. Swartz, The Vitali-Hahn-Saks Theorem for Algebras, J. Math. Anal. Appl., to appear.

18. S. Banach, Sur les operations dans les ensembles abstraits, Fund. Math. 3 (1922), 133-181.

19. S. Banach and H. Steinhaus, Sur le principe de la condensation de singularites, Fund. Math. 9 (1927), 50-61.

20. S. Banach, Theorie des Operations Lineaires, Warsaw, 1932.

21. C. Bessaga and A. Pelczynski, On Bases and Unconditional Convergence of Series in Banach Spaces, Studia Math., 17 (1958), 151-164.

22. Bourbaki, Espaces Vectoriels Topologiques, IV-V, Hermann, Paris, 1976.

23. J. Brooks, Sur les suites uniformement convergentes dans un espace de Banach, C. R. Acad. Sci., Paris, 274 (1974), A1037-1040.

24. J. Brooks and R. Jewett, On Finitely Additive Vector Measures, Proc. Nat. Acad. Sci., U.S.A., 67 (1970), 1294-1298.

25. J. Brooks and J. Mikusinski, On Some Theorems in Functional Analysis, Bull. Acad. Polon. Sci., 18 (1970), 151-155.

26. J. Burzyk, C. Klis, and Z. Lipecki, On Metrizable Abelian Groups with a Certain Summability Property, Colloq. Math., to appear.

27. J. Burzyk and P. Mikusinski, On Normability of Semigroups, Bull. Acad. Polon. Sci., 28 (1980), 33-35.

28. C. Constantinescu, On Nikodym's Boundedness Theorem, Libertas Math., 1 (1981), 51-73.

29. R. Darst, On a Theorem of Nikodym with Applications to Weak Convergence and von Neumann Algebras, Pacific J. Math., 23 (1967), 473-477.

30. M. Day, Normed Linear Spaces, Springer-Verlag, Berlin, 1973.

31. P. Dierolf, Theorems of the Orlicz-Pettis Type for Locally Convex Spaces, Manuscripta Math., 20 (1977), 73-94.

32. P. Dierolf and C. Swartz, Subfamily-summability for Precompact Operators and Continuous Vector-valued Functions, Rev. Roumane. Math. Pures. Appl., 26 (1981), 731-735.

33. J. Diestel and B. Faires, On Vector Measures, Trans. Amer. Math Soc., 198 (1974), 253-271.

34. J. Diestel and J. Uhl, Vector Measures, Amer. Math. Soc. Surveys #15, Providence, 1977.

35. J. Dieudonne, History of Functional Analysis, North-Holland, Amsterdam, 1981.

36. L. Drewnowski, Equivalence of Brooks-Jewett, Vatali-Han-Saks, and Nikodym Theorems, Bull. Acad. Polon. Sci., 20 (1972), 725-731.

37. L. Drewnowski, Uniform Boundedness Principle for Finitely Additive Vector Measures, Bull. Acad. Polon. Sci., 21 (1973), 115-118.

38. N. Dunford and J. Schwartz, Linear Operators I, Interscience, N. Y., 1958.

39. H. Hahn, Uber Folgen linearen Operationen, Monatsch. Fur Math. und Phys., 32 (1922), 1-88.

40. R. Haydon, A non-reflexive Grothendieck space that does not contain ℓ^∞, Israel J. M., 40 (1981), 65-73.

41. E. Hewitt and K. Stromberg, Real and Abstract Analysis, Springer-Verlag, N. Y., 1965.

42. T. H. Hildebrandt, On Uniform Limitedness of Sets of Functional Operations, Bull. Amer. Math. Soc., 29 (1923), 309-315.

43. J. Howard, The Comparison of an Unconditionally Converging Operator, Studia Math., 33 (1969), 295-298.

44. N. J. Kalton, Spaces of Compact Operators, Math. Ann., 208 (1974), 267-278.

45. C. Klis, An Example of Noncomplete Normed (K)-Space, Bull. Acad. Polon. Sci., 26 (1978), 415-420.

46. G. Kothe, Topological Vector Spaces I, Springer-Verlag, N. Y., 1969.

47. G, Kothe, Topological Vector Spaces II, Springer-Verlag, N. Y., 1979.

48. I. Labuda and Z. Lipecki, On Subseries Convergent Series and m-Quasi Bases in Topological Linear Spaces, Manuscripta Math., 38 (1982), 87-98.

49. J. Lindenstrauss and L. Tzafriri, Classical Banach Spaces I, Springer-Verlag, N. Y., 1977.

50. I. Maddox, Elements of Functional Analysis, Cambridge Univ. Press, 1970.

51. S. Mazur and W. Orlicz, Uber Folgen linearen Operationen, Studia Math., 4 (1933), 152-157.

52. S. Mazur and W. Orlicz, Sur les espaces metriques lineaires II, Studia Math., 13 (1953), 137-179.

53. J. Mikusinski, A Theorem on Vector Matrices and its Applications in Measure Theory and Functional Analysis, Bull. Acad. Polon. Sci., 18 (1970), 193-196.

54. J. Mikusinski, On a Theorem of Nikodym on Bounded Measures, Bull. Acad. Polon. Sci., 19 (1971), 441-443.

55. W. Orlicz, Beitrage zur Theorie der Orthogonalent wicklungen II, Studia Math., 1 (1929), 241-255.

56. E. Pap, Funkcionalna analiza, Institute of Mathematics, Novi Sad, 1982.

57. E. Pap, Contributions to Functional Analysis on Convergence Spaces, General Topology and its Relations to Modern Analysis and Algebra V, Prague, 1981.

58. A. Pelczynski, On Strictly Singular and Strictly Cosingular Operators, Bull. Acad. Polon. Sci., 13 (1965), 31-36.

59. B. J. Pettis, On Integration in Vector Spaces, Trans Amer. Math. Soc., 49 (1938), 277-304.

60. R. Phillips, On Linear Transformations, Trans. Amer. Math. Soc., 48 (1940), 516-541.

61. A Pietsch, Nukleare Lokalkonvexe Raume, Akademic-Verlag, Berlin, 1965.

62. A. P. Robertson, Unconditional Convergence and the Vitali-Hahn-Saks Theorem, Bull. Soc. Math., France, Suppl. Mem., 31-32 (1972), 335-341.

63. S. Rolewicz, Metric Linear Spaces, Polish Scien. Publ., Warsaw, 1972.

64. H. Rosenthal, On Complemented and Quasi-complemented Subspaces of Quotients of C(S) for Stonian S, Proc. Nat. Acad. Sci., U.S.A., 59 (1968), 361-364.

65. H. Rosenthal, On Relatively Disjoint Families of Measures with Some Applications to Banach Space Theory, Studia Math., 37 (1970), 13-36.

66. S. Saxon, Some normed barrelled spaces which are not Baire, Math. Ann., 209 (1974), 153-160.

67. W. Schachermayer, On Some classical Measure-Theoretic Theorems for non-sigma-complete Boolean Algebras, Dissert. Math., to appear.

68. J. Sember, On Summing Sequences of O's and 1's, Rocky Mt. J. Math., 11 (1981), 419-425.

69. C. Swartz, Applications of the Mikusinski Diagonal Theorem, Bull. Acad. Polon. Sci., 26 (1978), 421-424.

70. C. Swartz, A Generalized Orlicz-Pettis Theorem and Applications, Math. Zeit., 163 (1978), 283-290.

71. C. Swartz, Orlicz-Pettis Topologies in Function Spaces, Publ. de L'Inst. Math., 26 (1979), 289-292.

72. C. Swartz, A Lemma of Labuda as a Diagonal Theorem, Bull. Acad. Polon. Sci., 30 (1982), 493-497.

73. C. Swartz, The Schur Lemma for Bounded Multiplier Convergent Series, Math. Ann., 263 (1983), 283-288.

74. C. Swartz, An Abstract Orlicz-Pettis Theorem, Bull. Acad. Polon. Sci., to appear.

75. C. Swartz, Continuity and Hypocontinuity for Bilinear Maps, Math. Zeit., 186 (1984), 321-329.

76. C. Swartz, The Schur and Hahn Theorems for Operator Matrices, Rocky Mountain J. Math., to appear.

77. G. E. F. Thomas, L'integration par rapport a une mesure de Radon vectorielle, Ann. Inst. Fourier, 20 (1970), 55-191.

78. I. Tweddle, Unconditional Convergence and Vector-Valued Measures, J. London Math. Soc., 2 (1970), 603-610.

79. A. Wilansky, Modern Methods in Topological Vector Spaces, McGraw-Hill, N. Y., 1978.

80. K. Yosida, Functional Analysis, Springer-Verlag, N.Y., 1966.

INDEX

NOTATION